Through Seventeen Borders

JOHN H. DENENBERG

First Edition

PAGE PUBLISHING
Conneaut Lake, PA

First originally published by Page Publishing 2022

ISBN 979-8-88654-995-9 (pbk)
ISBN 979-8-88654-994-2 (digital)

Printed in the United States of America

Contents

The Bubble

Everyone has a time in their early life when they have their youth especially, no commitments and no responsibilities. This bubble in life breaks one day without notice and can never happen again. This is the true story of an adventure in Jack Dean's bubble.

This story is told from the recollections of Jack Dean, now a grandfather, over fifty years later.

The Beginning

It all began in the late sixties, when Jack Dean met Joseph Bly at a party. Joe was a couple of years older than Jack, but a friendship developed immediately and they began hanging out together all the time.

The war was raging in Vietnam. American kids, the same age as Jack, were being conscripted into the US Armed Forces and ended up in Vietnam, fighting a very unpopular war. Antiwar protests were commonplace. Young people rebelled against the Establishment and the conservative lifestyle of their parents. Lots of young people were smoking grass and hash. Many others experimented with mind-altering drugs. Pills to go "up" or go "down" were everywhere. It was the time of the hippie. It was the time for change.

True hippies broke away from society to live off the land with only the basics in life. They formed communes with like-minded others and, when possible, grew their own food, made their own clothes, and tried to live apart from the material world and the Establishment they put down. They lived a life of peace and harmony and love for their fellow man, but they had few possessions. Many others grew their hair long and believed in the peace-and-love movement but did not want to give up everything and live on a commune. Those hippies expressed themselves in other ways. They wrote songs about the war in Vietnam. They wrote songs about love, peace, and freedom. They protested for change. Music became the vehicle for expression.

Jack and Joe enjoyed being part of the revolution. They grew their hair and dressed the part, truly believing in peace and love. And sex and rock and roll too.

It was a sunny day in Montreal in July of 1969. Jack was sitting on his father's porch, covered in baby oil and taking in the sun with his sun reflector while listening to some tunes. The phone rang. It was his buddy Joe.

Joe had fantastic news. He and another friend, Moose, had planned to just take off and drive to the West Coast. This was in the planning stage long before he even met Jack. They decided to leave the next day for Berkeley, California, indefinitely. Jack's reaction was immediate. He was jealous. He wanted to go too. Joe said that once he got down there, he intended to find some work doing anything and secure a place to stay. Perhaps he could phone Jack and they could meet up in Berkeley when he was settled. Jack jumped all over that. Deal. Deal.

Berkeley was the epicentre of the love child and the hippie movement, just across the bridge from San Francisco. The University of California–Berkeley campus was one of the top ten research universities in the world. Timothy Leary got his PhD there. He was a legend in those days, an advocate of psychedelic drugs.

It was a huge decision for Joe to just pick up and drive four thousand miles to a different country without a thing in place. All he had was a half-baked plan and a burning desire for adventure. Perhaps more importantly, he believed in his heart that everything would work out. It always did. You just had to keep the faith. Only the young, before enduring the pitfalls that life had to offer, have the ability to believe without question. Jack and Joe both looked at life in the same positive way, a cup always being half-full and never half-empty.

Jack lived with his dad, Pops, and was between jobs, so he was free as a bird. The time was perfect, and he knew it. Jack began to prepare for his trip so he could leave the moment Joe called. He serviced his little mustard Fiat 850 convertible and ran around, getting some things he felt he needed for the trip, as if he were leaving the next day too. He knew it might be a couple of weeks before Joe called, but he wanted to be ready.

Joe left the next day with Moose, as planned. He called Jack to say "Adios" and confirmed that he would call as soon as they had found a place to stay.

Joe called in just over a week. Jack was quite surprised. It took at least a week to drive there, so they must have found a place right away. Joe told Jack that they had met some "good people" who knew of a place to stay. Joe gave Jack the address. In the sixties, "longhairs" used to flash a peace sign as another longhair approached, and if the person flashed one back, there was an immediate bond. And that was what had happened with their newfound friends, who got them the apartment. One of the guys had a job as a roofer and offered to ask his boss if he needed a couple of extra guys to work. His boss was hiring and agreed to take on Joe and Moose. And if it worked out, they could probably get Jack a job as well.

On the Road

Jack left for Berkeley the next morning. It was the first time he had ever left home on his own. It was thrilling to jump into his car and hit the open road without a care. He had an eight-track tape deck with tapes of *Led Zeppelin, the Beatles' Abbey Road, Steppenwolf, Bob Dylan, The Rolling Stones, Janis Joplin*, and many others.

The first day on the road went well. Jack planned on driving five hundred miles every day. Jack stopped off in little towns along the way to get a motel room, eat, and gas up for the next leg of the trip. Once he drove his planned five hundred miles, he kicked back and rested. That usually involved going to the closest bar and chatting up the locals. They were always thrilled that a stranger showed so much interest in their town. They bragged about their town, told all kinds of blown-up stories, and sometimes, when they got really drunk, even confided in Jack as to who was doing who. Jack had a lot of fun.

It took almost a week to get to Berkeley. There were long-haired people everywhere. Jack saw dozens of people hitchhiking to San Francisco on the main drag. It was all so exciting. He couldn't wait to explore, but first, he had to find Joe and Moose. He headed to the address Joe had given him. When he had filled up with gas earlier, he had also purchased a map of Berkeley. As a result, he had made it to his destination without asking a single person for help.

Jack parked his car in front of the apartment building. He pushed open the front door, which was semiajar, and entered a small vestibule. There were four mailboxes on the wall. Below the mailboxes stood a little table with stacks of mail that looked like previous tenants' mail or wrong addresses, return-to-sender type of stuff. There was a staircase leading up to the first floor of apartments. Jack climbed the stairs and saw apartment number 1. That was it. He was really excited but kept his composure and knocked on the door. Within a minute or so, the door opened and a pretty young lady stood there. Jack asked her if Joe was there. She didn't even answer; she lunged at Jack and gave him a superwarm hug while directing him inside. The girl's name was Sandy. Sandy explained that Joe and Moose were at work and they would be back by six. In the meantime, he was stuck with her. Jack thought to himself that he didn't mind being stuck with her one bit.

They drank coffee and talked most of the morning. She told Jack that Joe and Moose couldn't wait to see him. She also told Jack that the apartment was hers and she was just letting the boys stay there until they found something else. Now that Jack had arrived, they could chip in for a larger place. It surprised Jack that Joe didn't mention they would be looking for another place so soon, but Sandy was right; it did make sense. They all couldn't stay there. Joe probably didn't want to complicate things by telling Jack that their accommodations were very temporary.

At about four o'clock, Joe and Moose rolled in. There were hugs and backslapping as soon as they saw Jack. Everyone was talking at once. Sandy brought four beers from the kitchen and passed them out. They all relocated from the hallway to the living room, where they could sit down and talk about their plans. Right off the bat, Joe announced that he and Moose had been fired that afternoon. Jack knew Joe didn't have a clue about roofing, so he wasn't surprised. Jack actually chuckled to himself when Joe told him he was going to do some roofing. But it was all good. He and Moose hated the job, anyway. The boss worked them like dogs, and they came home filthy. Where they were going to stay was another matter. Sandy said she would let them stay there until they could find a place. Sandy was "good people." They all vowed to start looking for a new place to live and not stop until they did. Tomorrow!

They ordered in Chinese food and sat around drinking beer and smoking joints. When they were ready to pack it in for the night, sleeping arrangements were discussed. As the new guy on the block, Jack was willing to sleep anywhere he was told. It was all figured out, and everyone scurried off to their place for the night. Jack noticed Sandy was sleeping alone, so he knew Joe or Moose didn't have an intimate relationship with her. That was hard to believe. She was cute and funny, with a great personality and a body to kill for. Jack really liked her.

They all got up about nine o'clock the next morning. Sandy made coffee for everybody, and as they sat around, drinking their coffees, they began forming a plan of attack for finding their new digs. While the boys discussed what they were going to do, Sandy was in the kitchen, making scrambled eggs and heaps of toast. When breakfast was ready, Sandy called them to the table and they all sat down to eat.

Our New Digs

The first thing they did was get a newspaper.

Jack was elected to run to the corner to buy the newspaper. As soon as he returned, they jumped right in with their task. They turned to "Rooms for Rent," and an ad jumped off the page. Well, everything was meant to be, so that ad was circled. They systematically went through all the other ads, finding something wrong with each one. That made their next move simple. They called the number in the only ad they had circled. They made a two-o'clock appointment for that afternoon.

They set out early for their appointment so they could check out the neighbourhood. They didn't know the town yet and didn't want to end up in a rough area.

They found the house and were pleased to see it was very well maintained. So far, so good. All three of them swaggered up to the front door and knocked. No one answered. They knocked again, a bit louder this time. At last, a guy opened the door. He appeared to be a few years older than them, but not many. He looked them up and down and asked what they wanted. Jack answered that they had phoned earlier and made an appointment to rent the room. The guy nodded and invited them to come in.

The owner introduced himself as Pete. The boys introduced themselves, and they all shook hands. Pete led them up to the second floor into a room that was about twenty feet by twenty feet. Keeping in mind that any arrangements they made would be temporary, they were satisfied that they could all sleep in that room. They all had sleeping bags, and another friend had donated three Airfoam mattresses to their cause. From there, Pete showed them the bathroom, which was to the right of the bedroom. It was clean and had both a bathtub and a shower. They went to the other side of the bedroom and into a small kitchen with a fridge and a hot plate, no stove.

The space was perfect for their simple needs. It was time to see how much Pete wanted for rent. They pretended to discuss what they were going to do, knowing full well they would take it if the price was right. They didn't want Pete to think it was a lay-down before he gave them the price.

Jack told Pete that they were willing to move in, depending on the cost. Pete said $40 a month in cash for the run of the house, excepting the first floor, which was his place. They could bring friends over and even have parties if they liked, providing they were respectful of his property and were sure to invite him. Jack agreed with the price but asked if they could pay him $10 every week, considering it would be a temporary agreement. They promised to give him a week's notice before they left. The vibes were great from the start, so Pete agreed. They all shook hands, and a deal was made. The boys had a new place to lay their heads. Pete gave them a key to the front door, with a reminder that they had to be quiet if they came in late.

Now that they had secured a new place, they returned to Sandy's apartment to tell her the good news. She was thrilled for them. And also a little relieved, no doubt. Three guys staying with her in a one-bedroom apartment was a bit much, but it was a moot point now. Sandy grabbed four beers from the fridge, and they toasted to the boys' new place and to great things for everybody in the future. When they finished their beers, they said their goodbyes, with promises of getting together soon. Jack never saw Sandy again. He never forgot her either, for some reason.

After leaving Sandy's place, they went straight to their new home. When they got there, they grabbed all their things and brought them in the house. Pete heard them and came to see if they needed a hand. Pete turned out to be a prince of a guy. He seemed to really enjoy the boys being around. Anything they needed was no problem, and he helped them navigate around town, pointing out the best places to eat and where to go to have some fun. They met a lot of really good people because of Pete. The boys all thought he was a super cool dude.

Everywhere they went, they met friendly people looking to party. Everything was "Far out" and "Groovy, man." Berkeley was known as a huge party town. At any given time, there were dozens of people hitchhiking to San Francisco, as Jack witnessed when he first arrived. Hitchhiking was very popular everywhere in the sixties, even back home in Canada. But Jack had never seen so many people with their thumb out at the same time. There were a lot of stoned-out people walking around town, some even tripping on acid. Many women walked around braless and believed in free love and drugs, sex and rock and roll. The boys were newcomers to Berkeley, but so were many others.

A couple of weeks went by with some sort of get-together or a full-blown party just about every day. At one of these parties, a girl named Natasha was going on and on about a free Rolling Stones concert where there would be no police; the Hells Angels motorcycle gang would be the security. Jack thought it could be fun to see the Stones live. When Natasha said that the concert would be remembered as historic for years to come, Jack was sold. He asked Joe and Moose how they felt about going. They both said they wanted to go.

It turned out that Natasha, like many other people, thought the Hells Angels would keep control and act like the police. The Hells Angels never officially committed to policing the event. Mick Jagger had hired them to keep people off the stage and protect the equipment from getting damaged. And for doing only that, they were paid $500 in beer, which was a shitload of beer. The Grateful Dead and Jefferson Airplane had used the Angels at other events as security for the bands, and things had gone well. The Angels never agreed to act as the police. The gang hated the police almost as much as they hated rules, let alone enforcing them.

Jack told Natasha that they were interested in going along. She was thrilled. She was that type of high-energy person. They made arrangements to meet the next morning at eight o'clock. The plan was to camp out for one night, and the concert would start the following day. Natasha reminded them to bring their sleeping bags.

The Stones Free Concert

They got up early the next morning and packed up the cars. They had to go in two cars because both their cars were two-seaters and there were three of them. They all grabbed a coffee, but Moose was hungry, so he poured himself a big bowl of Frosted Flakes.

When Moose was finally done, they set out to meet up with Natasha. As they drove up, they saw Natasha was already at the meeting place. She was with some friends in another car. When she saw them, she jumped out of the car and ran up to them and gave hugs all around. She yelled out to follow them as she jumped back into her car.

Off they went in a four-car convoy on a trip that should have taken about forty-five minutes. After driving for over an hour, the lead car pulled over. The driver got out of his car with a shrug and informed everyone he was lost. As soon as the words were out of his mouth, an oncoming car driving on the other side of the road stopped and asked if they needed help. Jack yelled out that they were going to the Stones concert. So were they, and they motioned with an arm to follow them. So everyone turned around, and now they were a convoy of five. By this time, there were so many cars going to the same place that it felt like they were driving in rush hour.

They drove for about thirty minutes. They were going to Altamont Speedway between the city of Tracy and the city of Livermore. When they got there, they saw hundreds of cars, vans, and pickup trucks parked in this huge field. They all parked in a row and locked up the cars. The stage set up for the concert was about a quarter mile away from the cars, farther into the field. They began walking until they reached the stage. Hundreds of people were already stationed in front of the stage, claiming the best territory they felt they deserved for arriving so early. Jack's group headed farther back, where they had more room to lay out their pup tent and three or four lawn chairs that Natasha had brought.

People were arriving by the hordes. By the time the concert started the next afternoon, there would be a crowd of three hundred thousand people. Jack looked back to where the cars were parked and wondered how they would ever find their cars again in all the chaos. Cars were everywhere, blocking other cars with no regard. It was easy to see lots of people tripping on acid and others nude or seminude, feeling the love and dancing to music playing in their minds.

Off to the side of the stage were the Hells Angels. There were dozens and dozens of bikers with their women, maybe hundreds, standing in front of a long line of motorcycles. As Jack walked by, he was careful not to make eye contact, fearing one of them might think he was staring at him or his woman and take issue. Jack did snap a couple of great pictures of them from a safe distance, though. Jack had met several Hells Angels who seemed a lot nicer than he expected when they were by themselves, but even the not-so-bad guys could get ugly when they got together with the gang, especially when they were drinking heavily.

As night fell, thousands of people were still arriving and continued to do so all night long. As the crowd swelled by the thousands, it was awesome to witness, but overwhelming at the same time.

They partied all night until the concert finally started the next day. They were popping "uppers" all night long, which made them euphoric and wide awake. They heard Jimmy Hendrix; Blood, Sweat & Tears; the Who; Crosby, Stills, Nash & Young; the Band; Creedence Clearwater Revival; Ravi Shankar; Arlo Guthrie; Joan Biaz; Janis Joplin (Jack's favourite), and many more great bands, including the Rolling Stones, of course. Even though they camped quite far from the stage, they knew something was happening in front of them. On a couple of occasions, the music stopped in the middle of a song, so it was obvious something was going on. But no one cared. Everyone thought it was most likely an equipment malfunction of some sort, never imagining there could be any violence at a "Peace and Love" concert.

The Angels were getting really drunk. They were getting into fights with the crowd, as well as with some of the band members. It was great to be part of what was happening at the beginning of the concert, but as the love turned to something else and after two days of not sleeping and doing drugs, Jack was ready to split. Everyone in their group was ready to leave, which was great, because there were a lot more eyes looking for their cars. They all began winding their way through the tens of thousands of people as the Stones sang. They were just passing the stage as they headed to where they figured the cars were parked and noticed a commotion in front of the stage involving the Hells Angels. They quickened their pace to the cars, not wanting to get in the middle of any trouble. They also needed to find their cars and get out of there before hordes of people left at the same time.

There was a lot of violence. Many people were hurt from all the fighting. Jack heard many cars had been vandalized. Four people died. Jack later found out that a Hells Angel killed an eighteen-year-old kid, Meredith Hunter, who was very high on drugs. He pulled a gun and was prepared to use it when a Hells Angel tackled him and stabbed him several times. Apparently, he stabbed him while Mick Jagger was singing "Under My Thumb." Jack and the gang walked out while Jagger was singing that song. The commotion they had seen in front of the stage was that kid getting killed. The Angel was later arrested and charged with murder. He pleaded self-defence, and a judge agreed and he was released. A few months later, they made a documentary of the concert at Altamont called *Gimme Shelter*. Because they had filmed the concert, they captured what had happened in front of the stage with the Hells Angels.

The "Stones free concert" in Altamont was supposed to be Woodstock West. Woodstock had happened four months earlier on the East Coast and represented peace and love, a reflection of the times. The concert at Altamont became known as the end of the hippie era and the de facto conclusion of the late-sixties youth culture. Natasha's words that the concert would be remembered for years to come were prophetic, but for a different reason. Jack and

Joe ended up in San Francisco later on and attended the official mock funeral procession, casket and all, of the hippie. The hippie was dead.

They found their cars, and nobody had boxed them in. Jack thought that was a miracle, especially considering so many others were trapped. In another hour, it would be bedlam there. Off they went, clear of the nightmare they had dodged. Their group all went their separate ways after appropriate waves. Joe needed gas, so they stopped off at a Shell station in Livermore. There was no one there gassing up at that time. They were ahead of thousands who would be lined up for gas very soon. As the boys filled their tanks, the attendants manning the booths where you paid suddenly started to freak out. They locked up their booths, ran from the pumps into their office, and locked that door too. As the attendants ran past the boys, they yelled at them to leave. Not realizing what was happening, Jack tried to explain that they hadn't paid yet. The attendants motioned to them to leave in a desperate wave of their hands while they ran inside to safety. And then they heard it. There was a roar of hundreds of Harleys coming from over the hill, right in front of them. The boys hightailed it out of there to a safe distance and watched biker after biker fill up with gas and just take off without paying. The attendants stood there and watched too. No one dared get in their way. The boys waited to see what direction the bikes were headed and made their escape in the opposite direction.

Before all hell broke loose, when they first pulled up to the station, a woman asked Jack if he would give her a ride. She looked weathered and had obviously been around the block a few times. Jack liked her after just a minute of conversation and told her to jump in, thinking she would be game to party later on. The boys were in their early twenties, and with soaring levels of testosterone, pretty much anything to do with the opposite sex looked attractive and exciting. Jack guessed she was used and abused so badly by the bikers they no longer wanted her around. Or she could have had a fight with her biker "old man" and just took off. If that was the case, Jack didn't want to be with her when the biker found his "old lady." Once they got out of the area, Jack made a stupid excuse and told her nicely she had to get out. She did so without a fuss, thankfully. And they continued on their way home.

As soon as they got home, they all crashed and didn't get up until noon the next day. Jack was up first. He made a coffee and sat on the sofa in the living room, still groggy from the concert experience they'd had. He became deep in thought, feeling good that he had gone to the concert. He thought he would probably never see the Hells Angels in force again for the rest of his life. That was something special to see. There were lots of gangs back home, but he hardly ever saw them. Out West, there were lots more, and you saw them all the time. Not that Jack wanted to be a biker. It was exciting.

Moose got up next and went into the living room to say "Good morning" to Jack, startling him out of his deep thought. Jack asked Moose what he thought about the last few days. He echoed Jack's feelings, saying he was glad they had gone, but he was happy to be home. It was very taxing. Bad stuff had happened, but none of it really affected anybody except the people in front of the stage. Those poor bastards who got there early to get their prime location had ended up in the worst place possible. They heard noises in the kitchen and knew Joe had finally gotten up.

They had been living in Berkeley for several months now. They had made a lot of friends. Several of Pete's friends had become their friends as well. Everyone seemed to be intransient. No one had jobs they liked. No one had a career. Everyone worked to make enough money just to get by. No one thought about tomorrow. It was all "party hardy" and "have a good time." The boys loved getting together with all their friends. They fell into a group of really good people. Big deal—nobody had any aspirations. They were young and weren't ready to settle down. The whole lot would look back and cherish those memories when their lives became a lot more serious down the road.

One of their good friends met a girl named Alice. She had come to Berkeley with a girlfriend a few months earlier, and her girlfriend ended up meeting a guy and moving in with him. That left Alice on her own to fend for herself. Alice was invited to one of the get-togethers and was introduced to everyone. She was a pretty blonde who got along as well with the ladies as she did with the guys. Jack liked her but didn't make a move. Joe, on the other hand, latched onto her immediately and began a thing with her. It was hot and heavy for a while, but as time crept on, the relationship started to wane. They never fought, but the luster of the relationship had tarnished.

It was about the time that Jack and Joe began talking about packing up and hitting the road again. They discussed going down the coast of California and right into Tijuana, Mexico. It wasn't long before their talking turned into action. All three of them sat down and shared how they felt about moving on. It was obvious that Jack and Joe wanted to leave. Moose wanted to get a job and stay and live in Berkeley. He had no money and couldn't afford to travel. He really liked Berkeley and had a bunch of friends they made to keep him company.

It was decided. Jack and Joe would head out in the morning, leaving Moose in Berkeley. That night, they got together with everyone to say their goodbyes. Lots of toasts were made, and many promises to keep in touch. It was sad to say goodbye to their friends, some of whom had become pretty tight with the boys. Joe and Alice had ended their relationship amicably and said they would also try to keep in touch. They both knew that would be unlikely.

The next morning, Jack and Joe tried to get up early to get a good start. They had gone to bed much later than they had planned, so getting up early proved to be harder than they had imagined.

Jack got up first and made a coffee. Joe materialized about a half-hour later and made himself one too. They both sat in a dazed state, trying to get into the headspace of leaving Berkeley after months of living there. The transition from safe and secure to the unknown required a new state of mind.

The door to their bedroom opened, and in walked Moose. Jack knew he had been in the washroom, because he wasn't in his bed. What blew Jack's mind, Joe's too, was Alice standing right behind him. They were up all night, rapping. They discovered that they both secretly wanted to be together but had been too shy to say anything. Joe later told Jack that he was really pleased with the way things had worked out because he had felt a bit guilty about how he had handled his relationship with Alice. Now, he knew that Moose would take good care of her, much better than if she were still with him. Moose didn't do that well with the ladies, so to have a good-looking babe who truly cared for him meant a whole lot to everyone, especially Moose.

On the Road Again

Finally, they were ready to move out. The cars were loaded with all their stuff, and the tanks were full. All goodbyes were said. That was the last time Jack and Joe ever saw Moose again. He died of an infection in his mouth a few years later while down in Mexico.

Their plan was to head down the West Coast and into Mexico. From there, they would head eastish back to Montreal. This plan was, of course, the basic plan, with emphasis on "basic." They had even talked about where they wanted to go after they got back to Montreal. Europe was an option, and so was the Middle East. Maybe the Far East, to India. Nothing was off the table. One always had to be mindful of their surroundings when traveling, then and now. That would never change. Back in those days, Afghanistan was a popular destination for longhairs. Anyone coming back from there, without exception, raved about how wonderful the Afghan people treated those visiting their country and how kind they were. They brought you into their homes and felt honoured to be your host. It is far too dangerous for anyone to travel there now, but if it once was, then it can be again, hopefully.

Their first destination was San Francisco. They were there at the exact time they performed the mock funeral of the hippie. And of course, there were hundreds of long-hairs trying to act cool in the procession, so Jack and Joe fit right in. They met people from all over the States and a group that came from as far away as Australia. They walked with the crowd until people started to disperse. Even back then, Jack was aware that he was participating in something historic.

They decided to see the city a little before they continued on. After going to Fisherman's Wharf, they saw Alcatraz across the San Francisco Bay. The water between San Francisco and Alcatraz was shark infested, making escape from the prison by swimming very unlikely. They wandered around the city a bit, staying close to the cars. To walk in San Francisco could be very difficult with all those enormous hills. Before packing it in and heading out, they stopped at a restaurant and had a San Fran specialty of Bay shrimp and crusty, sour French bread. They were glad they decided to eat before continuing on. The Bay shrimps were delicious.

They drove south towards Big Sur, a rugged stretch of California's central coast between Carmel and San Simeon. It was bordered by the Santa Lucia mountains on the east and the Pacific Ocean on the west. They were traveling along Route 1. The road had many winding turns, steep cliffs, and gorgeous views of the coastline. It was fun just bombing along the highway in such a beautiful setting, top down, tunes blasting, pedal to the metal. Along the way, they got to see some whales. Jack and Joe had only ever seen whales on television, never live in nature. They pulled over and watched the whales for a while. They were huge but majestic as they jumped out of the water and splashed back down. They were awesome to watch. They stopped to grab a bite as they drove through Big Sur. As they walked up to the front entrance of the restaurant, they saw a sign that read, "No hippies allowed. No bare feet." Jack and Joe saw a bunch of people in the restaurant staring at them through the windows as they walked up to the door. The boys sensed trouble and turned around and headed back to their cars and took off. Jack snapped a picture of the sign in the window before he hightailed it out of there.

They finally got to Los Angeles. As they drove into the city, they saw an advertisement for the rock musical *Hair*. It was playing at the Aquarius Theater on Sunset Boulevard. *Hair* celebrated the hippie counterculture and sexual revolution of the 1960s. It was so popular it remained at the Aquarius for more than two years.

Jack and Joe had talked about seeing *Hair* one day. They would never get closer than they were then, so they decided it was time. They found the Aquarius Theater and purchased two tickets for the next day at noon. Jack hit it off with the ticket girl, so she gave them primo tickets, centre stage, in the second row.

They looked for a motel on Sunset Boulevard and found cheap accommodations. They decided to grab a bite to eat and pack it in early, weary from all the driving.

The next day, they saw *Hair*. They really enjoyed the performance. They felt like they were part of the revolution, with their hair flowing down below their shoulders. They also felt a little proud that not only did they get to see *Hair*, which was all the rage, but they saw it at the famous

Aquarius Theater in Los Angeles. For an East Coast kid, that was pretty cool.

When they left Los Angeles, they kept going south to San Diego. They had heard San Diego was known for having an ideal climate all year round and the most beautiful beaches. They had planned to pick one of those beaches and take a dip. But they didn't stick around very long, though. Tijuana was only seventeen miles away from San Diego. They abandoned their plan to go swimming and headed for the Mexican border.

Before long, they were in Tijuana. There were lots of shops selling all kinds of leather goods and authentic Mexican clothing at really good prices. Joe bought a guitar and a wide-brimmed black hat, a suede coat with tassels, and a bunch of interesting Mexican wood carvings. Jack bought a leather vest and a sombrero, the latter as a souvenir, not to wear. He also picked up a beautiful embroidered blanket and some other things that had caught his eye. Joe wore that hat and jacket the rest of the way back to Montreal. He looked like he was a cross between a Mexican and a frontiersman. The more outlandish one dressed, the cooler they were. Joe always had lousy taste, but anything was accepted in those days.

It was New Year's Eve the night they arrived in Tijuana. They knew it was around the corner, but the date had kind of crept up on them. They discovered that the time in Tijuana, Mexico, and California in the States had a one-hour time difference. In other words, they could celebrate New Year's Eve in Tijuana and jump over the border and celebrate it again in the States an hour later.

After shopping, they sat down on the patio of a bar to relax and people watch. As luck would have it, two *muchachas* sat at the table beside them. Everyone, tourists and locals alike, had started celebrating the start of the New Year festivities. The girls were extremely friendly, making conversation in broken English. So they eventually all moved to one table. Things were going along well until about a few minutes before midnight. Two Mexican guys walked into the bar and came over to their table and hugged the girls. The girls introduced them to Jack and Joe. Jack knew immediately that they had to leave. He was getting really bad, dangerous vibes from those guys. Jack looked at Joe and could tell he was thinking the same thing. The Mexican guys were pissed off that Jack and Joe were with two good-looking Mexican ladies that they knew, and they weren't with anyone.

The clock struck midnight, and the fireworks began. Jack and Joe watched the festivities for a while, which were quite impressive with the Mexican touch. When everyone was busy watching what was happening, the boys took the opportunity to sneak off. They jumped into their cars and took off, escaping across the border into the States. They both felt that they had dodged a bullet by leaving when they did. Those guys would have punched the shit out of them in order to get the girls. They were probably overjoyed when everyone realized Jack and Joe had disappeared.

They celebrated New Year's Eve for a second time in the first bar they passed once they had entered the States. It turned out to be a small local crowd that all knew one another. The people were super nice. The vibes in the bar were fantastic, so the boys planned to party with those people until everything had shut down. They had no place to sleep, though. They were far too drunk to jump in the car and start looking for a motel in the wee hours of the morning. That night could be one of those rare times when one had to sleep in their car. That was unless they could finagle a better arrangement somehow.

They did get a place to crash. The beautiful female bartender offered them two couches in her apartment above the bar. Once everyone started to leave, it became known—*wink wink*—that the boys were going to sleep in their cars. She took pity on them. The girl's name was Angel. She dropped everything she was doing to clean up and brought them right up to her place. She showed them the couches, pointed to the washroom, and turned to leave. Before she left, she gave each of them a blanket. She said she had to go back to the bar to finish cleaning up, and off she went. Pretty trusting, Jack thought.

The boys were asleep when Angel returned. The next morning was New Year's Day. They all got up about ten. Everyone was hungover and didn't want anything to eat, so they sat around, drinking coffee and talking until it was time for Angel to go back to work. She worked in a bar that was open every day, especially on holidays. She started at noon. They thanked Angel for saving them from sleeping in their cars and hit the road. Angel was definitely "good people."

They continued east to Tucson, Arizona. From Tucson, they drove south for sixty miles to the Mexican border and crossed again into the town of Nogales. It was fun to experience the Mexican charm and traditions. They did a little shopping in some funky stores but didn't buy much.

After a few hours of poking around, they moved out. They returned to the States and continued east through New Mexico.

Driving through New Mexico, the boys decided to pull over, take a break from driving, and smoke a joint. They pulled onto a road off the highway and parked. Joe had a joint prerolled, as usual, so he pulled it out and lit it up. They smoked the joint and were just talking when Jack noticed a sign on the side of the road, a little farther up from where they were standing. It read, "No trespassing. Military training grounds. Do not enter." Although somewhat uncharacteristic of them, they decided to ignore the signage. Normally, neither one of them would ever disobey a "No Trespassing" sign, let alone a military "No Trespassing" sign. Without hesitation, they jumped in their cars and headed down the unpaved road. What a "rush!"

The Family

They drove along the dusty road, with no idea where it would lead them. They knew they were on private property, because of the signage, but didn't think anyone was around to cause them any harm. They were looking for a decent place to pitch their tent and enjoy sleeping under the desert sky. The sunsets in the desert sky can be magnificent. They pictured themselves finding a place to set up, building a campfire, and watching the sun go down.

The road twisted and turned as they progressed along. There were more "PRIVATE PROPERTY—NO TRESPASSING" signs posted every mile or so. Again, the boys ignored the signage and kept trucking on to some unknown destination.

At last, the road came to a dead end. There was a building there with a sign reading, "RESTAURANT." Outside the front door leading into the restaurant was a dog, a black terrier, wearing a hat and sunglasses. The dog was sitting on a tree stump, so very still, as if it were a statue. It was very weird. The boys were reluctant to touch the dog for fear it would snap at them or, worse, bite them. They walked around the dog with caution, not knowing how it would react. They kept as far away from it as possible, as they approached the front door of the restaurant. The dog sat still, never giving them any reason to fear him. Matter of fact, they felt they could have patted him but didn't dare try.

Before they entered, Jack took out his camera and snapped a picture of the dog. No one at home would ever believe this dog without a picture as proof. As soon as they passed the dog without incident, they entered the restaurant. They wondered how a restaurant could survive if the road was closed to the public. Who eats there?

They sat down at a counter, which ran the length of the restaurant. There were about fifteen barstools, maybe more. Three booths to accommodate four to six people each were along another wall. This restaurant was capable of seating a fair number of guests, which made the situation even stranger. There was a man standing behind the counter, presumably waiting to take their order. Sitting a few stools down from Jack and Joe was a middle-aged man with two young hippie girls sitting beside him. Jack wondered how the hell they had gotten there. They had to be trespassing also. As soon as the boys were seated, the man behind the counter came up to Jack and asked him if he was a man or a woman, because of his long hair. It was a very strange question to ask, but Jack answered without any hint of provocation in his voice. Jack calmly pointed to his beard and asked him if a woman grew such a beautiful beard. The man behind the counter chuckled and abandoned his line of questioning. He asked them what they wanted to eat. There was a sign on the wall that read, "Special. Two burgers and an order of fries and a drink, $1.50." They each took the special. The man disappeared into the back. He was probably doing the cooking, too, as there was no evidence that there was anyone else in the kitchen.

Jack and Joe sat quietly, waiting for their food as they tried to figure out the strange situation they found themselves in. The dog outside, the vacant restaurant at the end of a dead-end road, and the hippies at the end of the counter were all a mystery. Suddenly, the guy with the hippie girls stood up and approached Jack and Joe. He introduced himself as Joseph Sage. He turned to the girls and motioned with his arm. "This is Eve, and this is Beth." Jack and Joe got up and introduced themselves, and they all shook hands. Joseph told Jack he thought his response to the cook's question when they had first walked in was handled very well. He pointed out that Jack hadn't taken offense and answered the cook very calmly. Complimenting Jack on how he handled the situation was strange, Jack thought, but it definitely broke the ice. In no time, they were all talking to each other as if they were old friends.

Jack took the opportunity to ask Joseph to answer the questions that were bothering him and Joe. He asked Joseph about the dog in front of the restaurant. Joseph laughed. That was his dog, named Blackie. The girls loved to dress him up, and the dog was so good-natured that he let them do whatever they wanted to him. Next, Jack wanted to know how the restaurant could survive if no one was allowed to be there, except the military. There were probably maneuvers coming through every day, Jack figured. He asked Joseph about the restaurant, and Joseph confirmed that the military did maneuvers twice a week, and that was enough to keep the restaurant afloat. If the soldiers had been there that day, they would have already sent Jack back to the highway, or worse. Joseph told Jack that sometimes the soldiers exercised with live ammunition. Jack wondered why Joseph was allowed to stay but was intimidated to ask.

The food arrived, and the boys dug right in. They hadn't eaten anything all day, and they were starving. They both polished off their meals in record time. They asked for their bill and paid it. As they turned to leave, Joseph invited them over to where he and the girls were staying. The boys needed to pitch their tent before it got dark, but they figured they had time. Once the sun went down in the desert, it got really dark, and it would have made the task of setting up their tent all that more difficult. They accepted Joseph's invitation. They were secretly hoping they were both going to get laid.

The boys left their cars parked in front of the restaurant, because Joseph said they were only a two-minute walk away. Soon, they saw this converted yellow school bus parked in the middle of the desert. Beth ran ahead and opened the door. Joseph directed Jack and Joe inside before he followed behind. As the boys entered, there were three cages affixed to the wall. They appeared to house some sort of wild animals, about the size of a fox. Jack had no idea what they were or why Joseph had them. Joseph didn't offer any explanation, and neither Jack nor Joe asked what the animals were for. As they walked farther into the bus, the whole middle section was a kitchen. It seemed to have everything needed to prepare meals, with a full-size fridge and a stove with an oven. Beside the stove was a long counter area to prepare the meals. The entire rear of the bus was a huge wall-to-wall bed, where all the family members slept together. Cupboards were jammed into any space available for their clothes and their supplies. The "family" consisted of Joseph, as the leader, two guys who were doing some family business in Tucson, two other girls with them, and Beth and Eve.

It was a time of many runaways, most running from horrible situations at home. They all needed someone to give them the love that their parents never did. It was sad how many children were abused mentally or physically and still are. They easily fell prey to some people taking advantage of their situation. Joseph seemed to be one of the good guys, but many other so-called family leaders of other families were abusers who were experts at reading people. The downtrodden had no chance against them. Unconditional love, if nothing else, was what all kids needed then, now and forever.

The girls hadn't spoken much when they first met in the restaurant. As they spent more time together, they began to loosen up and engage the boys in more conversation. Eve offered Jack and Joe a macrobiotic cookie that she had just baked. She had told the boys earlier that Joseph was capable of reading minds. Eve really believed that. Jack and Joe thought the cookies were terrible, to put it mildly. Jack remembered concentrating on saying to himself over and over in his mind that the cookies were delicious, just in case Joseph really could read minds. While Joe was talking to Joseph, the girls asked Jack if he wanted to join their family. They had already asked Joseph, and he had given them his blessing.

This was the real thing, free love with multiple women whenever he wanted. But there was always a price to pay for everything, and the price that time was way too high. Jack was wide-eyed and bushy-tailed and there for the adventure. He wasn't running away from home because no one loved him.

Jack thanked them for asking him to join the family. He could tell by the girls' seductive body language that they were willing to do anything Jack wanted to entice him to join. Jack told the girls that it was a huge decision, and he wanted the night to sleep on it. He would give them his answer at breakfast in the morning. The girls were really happy Jack was considering joining them. Jack already knew his answer. There was no way and no reason to join their family. The family were mostly young kids that were down and out and searching for love. They wanted to belong to a real family. Jack tried to never judge others, but that was the way it was. Dusk started to provide a beautiful sunset, and they knew that nightfall wasn't far away. They still had to pitch their tent and prepare for the night ahead. So off they went. Now they were in a hurry.

The boys found a suitable place to set up the tent. They worked quickly to finish before dark. They just made it. Just as they finished, it was so dark that they could hardly see each other. They didn't bother with a campfire, just straight into their sleeping bags. As they lay there, Joe asked Jack if he had been asked to join Joseph Sage's family. Joe said he had overheard the girls ask Jack to join but didn't hear his answer. Jack told Joe that they did ask him to join and left it at that, forcing Joe to ask what his answer was, dreading what he would hear. Jack told Joe there was no way he would leave his *amigo*. Joe let out a sigh of relief, and they both fell asleep instantly.

Morning came, and they struggled to get up and dismantle their tent. They got it all done, threw all their stuff in the cars, and headed to the restaurant for breakfast. Jack thought he had better give Joseph and the girls his answer before they ate and get it out of the way.

They boys walked into the restaurant to find Joseph and the girls already there. Jack walked straight up to Joseph and told him, and everyone else, that he was flattered to have been asked to join their family, but he had made plans to travel to Europe. Nothing could ever change his mind. No pussyfooting around. There was some moaning and groaning, but not too much. It was a great excuse and was even true. Joseph had a disappointed look on his face. It was much more serious for him than for Jack. They thought they had almost found another male to increase their family numbers, which was what they were always trying to do. Jack never had to struggle with any decision. He would have told them his answer on the spot, but he thought that might have offended them.

And that was that. They all ate breakfast and said their goodbyes. It was only after Jack returned to Montreal that Joseph Sage crossed paths with Jack again, in a sense. Jack was reading the book, *Helter Skelter*. It was a book about the Manson murders, which had just happened several months earlier, the time Jack and Joe were out west. Linda Kasabian was a Manson family member who participated in the Sharon Tate and Bianca murders under Charles Manson's direction. After the gory murders, she was understandably freaked out and went to see Joseph Sage for advice. Joseph Sage feared any Manson retaliation after hearing what had happened and now knew what

Manson was capable of doing. That was the reason Joseph was "lying low" in the desert when he bumped into Jack and Joe. Linda testified against everyone, and they all went to jail for life or for many years. Charles Manson went to jail for life even though he never physically did any of the killings himself. Linda's own charges were dropped completely, and she went free in return for her testifying against the others. Jack and Joe were that close to all that but never knew it until they read the book once they returned home.

They headed south and crossed the border into Mexico, yet again, in the town of Juarez. It was a border town similar to the others they saw, with stores catering to the tourists. They never ventured farther into Mexico than the border towns. It was a taste of a different culture that appealed to them.

Jack needed some air in one of his tires, so they found a place that serviced cars. They decided to fix the tire and get oil changes for both cars at the same time. Jack was a firm believer in maintaining his car. As a result, he had fewer car problems and much less hassles. While the work was getting done by the owner, they all joked around and became relatively friendly. The subject of marijuana came up somehow. The guy asked if they needed some. He said he had some great stuff at a very low price. The boys had very little left of their stash and were interested in a good deal. The cost was ten dollars an ounce.

Once the work on the cars was completed, their new friend, Carlos, arranged to get two ounces for the boys. They had to leave their cars parked at the shop and jumped into a car with Carlos and his friend. Carlos was in the passenger seat, and his friend was driving. Jack and Joe sat in the back seat. The sun had just set as they drove into the pitch-black desert, not a light to be seen except for their headlights. Jack looked at Joe with concern in his eyes. They didn't even know these people. How could they have allowed themselves to land in such a vulnerable position? All of a sudden, the car stopped. Carlos talked to his friend in Spanish for a minute. Then he turned around to the boys and said he would get the grass, but he had to go alone. His friend didn't want any strangers in his house. They now saw a light from a house in the distance in the middle of nowhere. Carlos said the price was ten dollars an ounce and put his hand out for twenty smackers. Jack had no choice but to give it to him. Jack had to pull out his wallet and take out two ten-dollar bills and then handed them to Carlos. When Jack took the bills out of his wallet, he noticed that Carlos was checking out how much more money Jack had left. Carlos took off towards the house on foot and disappeared into the blackness in seconds. The driver didn't speak English, so there was no conversation. They all sat there in total silence. Jack figured if he and Joe were going to be killed or robbed, they would have done so already. Hopefully, Carlos was actually going to return and give them their grass.

After an agonizing hour, Carlos appeared out of the night and reported that his buddy was expecting a shipment of grass the next day. As a favor to Carlos, he was willing to sell the boys only one ounce now. Jack said fine. Carlos returned ten dollars and the car turned around, and soon they were back at the shop and their cars. They thanked Carlos and headed north towards the border and back into the States. The time they waited for Carlos in the desert was a nightmare. Anything could have happened. They were still "rushing" after that experience and made a pact to be more careful from then on. At least they had their well-deserved ounce of grass.

They crossed the border from Mexico into the state of Texas. Jack wanted to see the Alamo, so they drove to San Antonio. When Jack was a little whippersnapper, his hero was Davy Crockett. He wore a genuine "coon-tailed" Davy Crockett hat and carried an official Davy Crockett rifle. Davy Crockett was a real person, a fact not known to many. He was a frontiersman, a congressman, and a storyteller. He was known as the King of the Wild Frontier. His adventures, both real and fictitious, earned him folk hero status. The Alamo, where he died fighting for Texan independence from Mexico in 1836, became a museum. It brought back great memories of his childhood when Jack saw the Alamo in person. It was exactly how he remembered it when he watched *Walt Disney Presents* from "Frontier Land" on the family black-and-white television set.

They headed north to Austin, Texas. They met some longhairs there who invited them back to their house. Those longhairs all carried sawed-off shotguns and told stories of *them* against the *rednecks*, who also carried guns. Jack had noticed many a pickup truck with a gun rack in the cab and a shotgun openly displayed. This was all very foreign to Jack and Joe. They lived in Canada, where no one carried a gun except those in law enforcement or the few who worked in jobs that required a gun for protection. They took pictures with the group, all brandishing their weapons, with Jack and Joe sitting front and center. They spent a couple of days with those people, mainly because

Jack and Joe became quite enamored with a couple of their girlfriends.

The boys were anxious to get home. They decided to drive straight through to Kent State University, where students were holding huge antiwar rallies. It was all over the news. It would be a hop, a skip, and a jump to get home from there.

Kent State and Home

It was over twelve hundred miles from Austin, Texas, to Kent, Ohio. The boys traveled through Arkansas, Tennessee, Kentucky, Indiana, and into Ohio. It was a marathon of driving. The first thing they did when they got to Kent was to get a motel room and crash. They were exhausted. Tomorrow was another day.

They got up early after some much-needed sleep. They had a leisurely breakfast and then headed to Kent State University, the scene of the protesting students. The students were protesting further involvement of the United States into neutral Cambodia. The North Vietnamese used the Ho Chi Minh trail as a supply route, and it went through Cambodia. So President Richard Nixon decided to invade Cambodia to break this supply route. The entire country was up in arms over that decision.

Only months after Jack and Joe were at Kent State, rioting in downtown Kent was out of control, so the National Guard was called in. They pushed the protesters out of downtown and to the university campus, coming face-to-face with protesters on the Commons. The Commons was a grassy knoll in the centre of campus traditionally used as a gathering place for rallies or protests. At some point, the

National Guard fired sixty-seven rounds of live ammunition into the protesters, killing four students and wounding nine others. The whole thing took thirteen seconds. No group or individual had ever been held accountable. That incident is referred to as the Kent State massacre or the May 4 massacre. It really hit home for Jack and Joe because it all happened a few months after Jack and Joe stood in that exact place on the Commons. It marked the first time in US history that a student had been killed in an antiwar rally. Hundreds of universities, colleges, and high schools organized walkouts in protest, creating the largest student strike in history, up until that time. The student strike of 1970 further affected public opinion over the role of the United States in the Vietnam War.

The boys were at Kent State about four months before the students were killed. They had gone to Kent State because it was on the way home, and they had wanted to be at the peace rally. Little did they know, that rally would evolve into a horrible nightmare some months later.

The boys left and headed straight home. It was approximately six hundred miles to Montreal from Kent. They figured they would roll into Montreal around midnight, considering the multiple stops they liked to make. They decided to phone their buddy Munroe when they got to Montreal and go right to his place to crash. After being on the road for several months, Jack and Joe couldn't arrive at their parents' places in the middle of the night.

Iceland

The boys finally made it home to Montreal, the city where they both grew up. They were very proud to call themselves Canadians first, and Montrealers a very close second. They began looking for the best deal they could find to get them to Europe. The exact destination wasn't important. The cost was a huge factor. Matter of fact, it was the biggest factor. They wanted to experience different cultures and learn different languages in faraway places, but for the least amount of moola. They each scraped together about twelve hundred dollars, which was not an easy task. They needed as much money as they could get. Now they were traveling overseas for an indefinite length of time.

They found a super deal to Iceland on Icelandic Airways, leaving two days later. Just enough time to pack and say hello and goodbye to family and friends. The flight would land in the Icelandic capital, Reykjavik. There was an option of staying over one day to sightsee and then continuing on the following day to Luxembourg without any additional cost. The boys opted for the one-day layover, wanting to visit the natural hot springs. They booked it.

Iceland is "the land of the midnight sun." Jack never realized it, but any place above the Artic Circle was referred to as the "land of the midnight sun," not just Iceland. There were always advertisements that referred to Iceland as the land of the midnight sun. Greenland, Norway, and Alaska are just three other examples of where the sun shines at midnight. Jack never thought about it before, but of course, it made perfect sense.

The flight to Iceland was exciting. They were on the road again, this time to another continent. It was January, the middle of winter in Reykjavik. It started to get light about eleven in the morning, and the sun was down by four in the afternoon. Jack thought it must be incredible in the summer when some people were in a lawn chair, tanning in the wee hours of the morning, and others were fast asleep, having to get up for work in the morning.

The plane landed at 6:00 a.m., and they were checked in to their hotel by 9:00 a.m. It was still dark outside. The official language of Iceland is Icelandic, but just about everyone spoke English fluently. They inquired about the hot springs and were pleasantly surprised that the hotel provided a free shuttle bus twice a day to and from the hot springs. The first bus was leaving at 10:00 a.m., less than an hour wait.

They grabbed their bathing suits and went looking for some food before they left. A few minutes to 10:00 a.m., the boys were in line to board the bus. At exactly 10:00 a.m., the doors to the bus opened, and they boarded with three other hotel guests who only spoke Spanish. Other than the odd word and lots of sign language, not much was said.

They arrived at the hot springs and found the changing rooms. It was quite mild outside with a jacket, but with only a bathing suit and towel, it was rather brisk. Once they got into the water, it was heavenly. Jack's body was submerged in the water, leaving only his head exposed to the fresh winter air, yet he was toasty warm. The boys soaked there for over an hour.

They jumped on the hotel shuttle and returned to the hotel. They had endured a six-hour flight to Iceland and then went straight to the hot springs. They were tired, but instead of crashing, they decided to take advantage of the few hours of sunlight left in the day and find a nice place to eat.

They found a place to eat that displayed pictures of some of the house specialities. They all looked great, especially to someone who was really hungry. They each pointed to the one they liked and then pointed to a Coke. Coke was almost everywhere in the world. The girl scribbled down their order and left. The boys found a table and waited for their food. When the food arrived, their meals tasted as good as the pictures looked. The boys didn't talk to anyone, as they usually did. They just sat there, stuffing their faces in silence.

They returned to the hotel without any desire to check out the nightlife. The Icelanders are accustomed to long days of darkness or sunlight throughout the night, but it is definitely an adjustment for people visiting the country. They were exhausted and needed a good night's rest so they could get up early the next morning and have breakfast before going to the airport.

The bus pulled into the hotel about nine o'clock the next morning, giving all those who had opted for the layover plenty of time to board. At almost 10:00 a.m., the girl from the hotel checked the list of guests who were supposed to board the bus to the airport. She was shocked to discover that Jack and Joe were AWOL. She alerted her manager immediately.

The manager rushed up to the boys' room to warn them that the bus to the airport was about to leave. He knocked on their door with a sense of urgency. The knock woke Jack, and he opened the door. He was startled to see the manager. The manager informed Jack that the bus was about to leave for the airport. Jack looked around at Joe, who was still sound asleep. Jack looked back at the manager, thanked him for being so kind as to wake them, and told him to let the bus go. They would take a taxi to the airport. It was 10:00 a.m., but the sun had not risen yet, and for Jack and Joe, it felt like 4:00 a.m.

Jack woke Joe up. He explained that they had slept in and missed the bus to the airport. Joe leaped out of bed in a panic. They both showered in twenty minutes. They packed up their stuff and raced to the lobby. There was a taxi right out in front of the hotel, waiting for a fare. They jumped in and told the driver they had missed the bus to the airport. The driver said, "No problem," and booted it towards the airport.

They hadn't gone far when it started to rain. It was coming down in torrents. The driver couldn't see the road and was only able to crawl forward at about twenty miles an hour. All of a sudden, there was a loud bang and the taxi swerved to the left, and the driver compensated by turning the wheel to the right and then onto the shoulder to a full stop. He got out in the blinding rain to investigate what had happened. A few seconds later, he jumped back into the car and reported he had blown a tire. He assured the boys that he would change the tire and get them to the airport on time. With that, he jumped out again and began to work in the pouring rain.

As the boys watched the driver breaking his ass as they looked through the window, Joe pulled out a joint he had prerolled. With a little bit of a smirk, he lit it up. They passed it back and forth until it was done. The car was filled with smoke. They couldn't open any windows for ventilation because of the pouring rain.

The driver got back into the taxi once the tire was replaced. Rain water was still dripping off his hair and onto his shoulders. He was drenched through and through. He said nothing. The taxi driver could have thought the boys smoked a Turkish cigarette, or he knew they had smoked hash and didn't care. By the time they got to the airport, the driver had to have been high from the thick smoke still in the taxi. It was raining so hard it was impossible to open a window, even a crack. They paid the driver and included a handsome tip, considering the nightmare this poor guy had just endured.

They rushed to check in for their flight to Luxembourg. Luxembourg is a tiny country landlocked between Belgium, France, and Germany. It is the second-richest country, per capita, in the world, and the wealthiest in the European Union. They have three official languages: Luxembourgish, French, and German. Prostitution is legal in the many brothels throughout the city and in and around the red-light district. Jack and Joe had never been to a brothel and decided to check one out when they got there.

They landed in Luxembourg and went straight to their hotel. They were starving, so after getting settled in at the hotel, they headed out for dinner. They found a restaurant down a side street that catered to the locals. When the waitress came to their table to take their order, Jack explained to her that they were tourists. She giggled, as if it wasn't obvious. Jack wanted to try something typical of what a local might order. The waitress suggested *Judd mat Gaardebounen*, a dish of smoked collar of pork with blanched broad beans. The pork is soaked overnight, then boiled with all kinds of vegetables and spices until the meat is tender. It is covered in a creamy sauce and served with sides of boiled potatoes and beans. That dish is considered to be the national dish of Luxembourg.

They ordered two *Judd mat Gaardebounen* with two *Funck-Bricher* beers, a very popular beer with the Luxembourgers. When they finished their meal, they called the waitress over and informed her that her suggestion was excellent; they had really enjoyed the meal. She thanked them proudly. They paid the bill. Jack took a deep breath and asked her if she could refer them to a house of ill repute. Jack thought it was a bit kinky to be talking to this strange girl about going to a brothel to get laid, but it was accepted differently there because it was legal. Jack still got a tinge of excitement when he asked her because it felt like he was talking "dirty" to her.

The waitress smiled and proceeded to give them directions to a brothel not far from the restaurant. She, of course, never went there, but she had heard it was one of the best. Jack wondered how one would judge one brothel better than another. Maybe one had better girls, somehow.

Maybe one had nicer decor. Or maybe one was clean, and the other not so much. They thanked the waitress and set out following the instructions they had been given.

They arrived at an elaborately carved, oversize door with a doorman waiting eagerly to greet patrons. The doorman held the door open with a welcoming smile, and the boys went in. As soon as they entered, a beautiful young lady came up to them and introduced herself as Tammy. The boys each mumbled their names. Tammy inquired if this was their first time there. Jack didn't know if she meant the first time in that particular brothel or first time ever in a brothel. He nodded yes because it was a yes either way. Tammy brought them to a booth with a table, much like a maître d' would do in a restaurant, except you didn't order dinner.

The boys sat there, enjoying the relaxed ambiance, not knowing what would happen next. They were both nervous. They saw a single guy at another table, sitting there, drinking a beer as he waited. He was there when they came in. Another table had three guys laughing and joking around, out for a good time. There was another table with two guys wearing suits, probably releasing some stress after work.

A sweet young girl came to their table to take drink orders. They ordered two Heinekens. She turned and disappeared.

The beers arrived in minutes. Jack and Joe always spouted that they would never pay for sex, and yet there they sat, in a brothel in Luxembourg. They considered getting up and leaving but decided that it would be foolish to pass up the opportunity of being with a professional who was dedicated to the pleasure of men. They abandoned their "no pay for sex" policy and waited for their turn to proceed. Each of the other tables were filled with patrons who came in after Jack and Joe. The boys were next.

Jack was filled with anticipation and excitement now. Joe too. Another lady came up to their table and introduced herself as Polly. The boys introduced themselves. She asked them to follow her. They did. She led them into another room with over a dozen ladies of every colour and description. All the boys had to do was pick the girl of their dreams. And they did.

That night was an experience Jack and Joe would always remember.

They left the brothel with a smile on their faces and returned to the hotel. They both showered and hit the hay, boom, boom, boom.

They got up the next morning feeling great. They had both slept like babies. The plan was to grab some breakfast, which the hotel offered, and then head to the train station and get on a train to Cologne, Germany.

By the time they arrived in the lobby, all the tables were taken except one. There was a table for two which was butted up to another table for two people. The other table beside theirs was occupied by this cool-looking couple. Jack and Joe sat down immediately, not wanting to lose the last table available.

The couple next to them was really friendly. They had just returned from Amsterdam. They said it was only a three- or four-hour train ride from there, and it would be well worth it if the boys could squeeze it in. When the couple said they had visited Anne Frank's attic, that clinched it.

Both Jack and Joe had read the book *The Diary of a Young Girl* by Anne Frank.

They really wanted to see the attic where Anne and her family and four others had hidden from the Nazis for over two years during World War 2. Not to mention the canals and everything else Amsterdam had to offer.

They decided to go to Amsterdam. They only had a couple of days to spare before they had to be in Venice to board their ship to Haifa, Israel.

They got on the train to Amsterdam. When they arrived, they found a hostel not far from the train station. They got settled and set out to explore. As luck would have it, they met two very nice Canadian ladies from a small town in the Laurentians, north of Montreal. The girls were backpacking across Europe after graduating from nursing college. Amsterdam was their last stop, and they were flying back to Montreal in the morning. Their names were both Elizabeth, but one was called Liz, and the other Beth. Liz and Jack hit it off from minute one, and so did Beth and Joe. The girls were happy to bump into two Canadian guys whom they could party with on their last night in Amsterdam. The girls had been there for a week and had discovered some great places to eat at and things to see and do. So the girls led, and the boys followed. They brought Jack and Joe to their favourite restaurant. It wasn't fine dining, but the food was fantastic. And the people were warm and inviting. Jack could clearly understand why the girls had brought them there.

When they first sat down, they ordered a carafe of wine, which didn't last long. So they ordered another. In no time, they were on their third. They talked and laughed and got along as if they had been traveling together for months. The third carafe was polished off, and they had to decide on a fourth or get going.

It was getting late, and the girls had to get up early. Jack and Joe offered to escort the girls back to their hostel and then return to their own. The girls accepted the gallant offer. When they arrived at the girls' hostel, Jack couldn't believe his luck. It was their hostel too. When everyone realized they were all staying there, they paired off, and Jack and Liz went to the girls' room and Joe and Beth stayed in theirs. They all had a wonderful time. The girls left for the airport the next morning, with promises of hooking up back home. They only lived an hour drive from Jack back in Montreal.

After saying goodbye to the girls, Jack and Joe headed to the Anne Frank House. When they arrived, they saw the shelves that hid the entrance to the attic. The Gestapo must have passed those shelves several times with Anne and her family keeping perfectly still so as not to make a sound. How frightening! They also saw the attic where Anne; her mother, Edith, and her father, Otto; her sister, Margo; Herman Van Pels and his wife, Augusta, and their son, Peter; and Fritz Pleffer hid for over two years, eight of them in total. They were completely dependent on six helpers, employees, and friends of Anne's dad, Otto. The helpers provided them with food and clothing, as well as books, magazines, and newspapers. All six helpers risked getting shot on the spot by the Germans if they were discovered, every single day.

They saw the skylight Anne had looked through, where she had watched the birds flying freely and had yearned for the war to be over. The day Anne and her family dreaded came without notice. No one ever found out if someone had informed on them or if it was just a fluke that they were discovered. They were huddled together by the Gestapo in the living area and relieved of all valuables, both on them and anything hidden away. The Gestapo rummaged through all their possessions, breaking anything they thought might be hiding something of value. Then, they were all hauled away and sent to concentration camps.

Anne and her sister, Margo, were shipped to the Bergen-Belsen concentration camp, where they both died of typhus only weeks before the Allied forces liberated the camps. They were both buried in a mass grave right at Bergen-Belsen. Everyone else from the attic died in the gas chambers, except for Anne's father, Otto. He was the lone survivor. A woman cleaning the attic after the group was arrested found Anne's diary and kept it so she could return it to Anne after the war. Unfortunately, she was never able to do so and gave it instead to Anne's father, Otto.

The boys left there teary-eyed and went for lunch along the canal. The canals were busy with activity. There were houseboats, some for rent and some that people lived on. There were boats packed with tourists cruising the waterways. It was a beautiful, sunny day, and everyone was out and about. Jack spotted an outdoor patio where they could have a few beers and watch the action on the canal.

It was a few hours and several beers later when they realized they were running out of time. They said goodbye to the people they had met on the patio, and down the hatch went their beers. The red-light district was next. They just wanted to see it. But then again, they were always open to changing course. It was a memorable time, hoisting a few alongside the canal in Amsterdam on a beautiful day.

They walked through a couple of bazaars and made it to the red-light district. You name it, they had it. A guy tried to sell Jack a ticket for five dollars to see a live show of a guy screwing a woman in twenty-five different positions. Live on a stage, as if you were in a movie theatre. They strolled down a walkway with glass-front brothels. Girls sat or stood in a window, allowing customers to see what they were getting. If you liked what you saw, you went in. If you walked by a window and nobody was there, it meant she was busy with someone else in the back. Each customer only got twenty minutes—that was it. Unless they weren't busy and they liked you. Other than a break between six and eight in the morning, those girls were open for business.

Rosse Buurt is what the people of Amsterdam call what is actually three red-light districts. The boys were in *De Wallen*, the oldest and largest area. The two other prostitution areas were Singelgebied and Ruysdaelkade, all three making up the *Rosse Buurt*. After a while, the boys had seen enough. It was back on a train in the morning. At 7:00 a.m., they would leave for Cologne, Germany. From Cologne, it was only four hours to Venice, Italy. That was where their ship was docked.

Our Ship

They boarded the train. The train had an aisleway with compartments to sit in on one side. The other side of the walkway were the windows of the train. Each compartment held a maximum of eight people, and each compartment had a sliding door. If the compartment was full, you slid the door shut. If there was still room to sit down, the door should be left open so that people could see the seat or seats available. Jack slid the door shut, hoping to deter anybody from entering so they would have the compartment for themselves. If anyone knocked, he would have to let them come in and sit down. And people did knock, saw a spot, and sat down. If anyone farted, the smell lingered for a long time.

Four hours passed quickly. They both slept the whole way.

Venice was busy. In order to get to the hotel, like everything else in Venice, you had to purchase a ticket for a water taxi. There were lineups. Once the taxi dropped them off, it was a struggle schlepping all their luggage to their hotel. Jack and Joe were dropped off at a wharf just about a block from their hotel, so it wasn't too bad. Some tourists had a lot farther to go to get to their hotel and had to pay someone to help them.

They checked into their hotel. It was beautiful. It wasn't like a hotel in North America. It was a house, with a sitting room outside the bedroom. There was a little garden area where you could sit and have a beer or a cup of tea in the afternoon or evening. Everything had that European flair.

Jack and Joe got settled in the hotel and then ventured out to see Venice. They headed to the central square, Piazza San Marco. They were hungry and chose one of the many restaurants with outside seating. They ordered a couple of beers before they ordered their food and just sat there watching the many tourists and the sights in the Piazza San Marco.

St. Mark's Campanile and Basilica stood at one end of the square. They noted that St. Mark's Campanile would be on the list of things to see the next day. At that moment, they were content to people-watch and check out what they wanted to see the next day, the day before they boarded their ship to Israel. The basilica was tiled with Byzantine mosaics, and one could climb the *campanile* tower and get a view of all of Venice. Venice was built on more than one hundred islands in a lagoon in the Adriatic Sea. There were no roads, just canals. The canals were linked by over four hundred bridges. Many times, Venice had been ranked the most beautiful city in the world. It was the birthplace of Antonio Vivaldi, regarded as one of the greatest Baroque composers.

Gondoliers trained rigorously and were considered one of the most highly regarded and most sought-after professions. They made a ton of money by anyone's standards. There was nothing more romantic than a couple on a gondola cruising down a canal, with the gondolier singing a love song. If you stood on one of the hundreds of bridges traversing the canals, you could watch the gondolas with their lovebirds and listen to the gondoliers sing. It was storybook stuff.

Jack and Joe were ready to eat. They saw what looked like a chicken dish at the next table, and it looked really good, so when the waiter came back, they ordered what the next table had and a couple more beers. They were perfectly happy to sit there all night, drinking beer and talking to other tourists. And so that was what they did.

The next morning, after breakfast at the hotel, the boys headed out to Piazza San Marco. It was packed with tourists. There was a lot to see right there. They went through St. Mark's Basilica, climbed the bell tower, saw Doge's Palace, and began walking through the city, entering any interesting place they found. They watched the main waterway, which supported everything that happened in Venice. Every restaurant and store was supplied by goods coming by boat or barge. Garbage was picked up by boat. Everything coming or going had to go from the wharf or to the wharf. It was amazing.

It was enough for one day. They were bushed. Earlier, they had seen a restaurant right near their hotel that they had wanted to try. They headed back in what seemed like the right direction, along alleyways they didn't recognize. The city was small enough that they knew they would eventually reach somewhere familiar. The walk back was so beautiful—the smells, the architecture, the European in everything. They arrived at the restaurant and had a meal that was better than they had expected. They returned to their hotel and were out like a light in no time.

Seven in the morning came quickly. They had an hour to shower, get out, and hit a fruit stand for breakfast before arriving at the dock by 8:00 a.m. When they arrived at the location, they saw their ship. It wasn't a very large ship. It had three levels, four if you counted steerage, belowdecks. Jack and Joe were in the steerage class. The term *steerage* originally referred to the part of the ship belowdecks where the steering apparatus was located. Over time, the term came to refer to the part of the passenger ship belowdecks where third-class passengers were housed, and therefore the cheapest cost for a ticket. Jack and Joe had paid only one hundred dollars each for five and a half days on the Mediterranean, so they didn't expect much. All the people in steerage were young kids with little money and looked at the trip as an adventure and a way to get to Israel cheaply.

Up the gangplank went Jack. Joe was right behind him. A gentleman wearing a uniform greeted them and asked to see their tickets. He glanced at their tickets and directed the boys to a door leading belowdecks. He motioned to a sailor standing there, waiting for his instructions, to show the boys where the men's dormitory was. The sailor walked them through the doorway, down some stairs, and into the hallway with several doors. The door to the left said "Men," and the door on the right said "Women."

The guy helping the boys was named Gino. He was about their age, and they got good vibes from him. He led them through the door that said "Men," of course. The room had about a dozen bunk beds. Nothing else. They had to put their luggage underneath the bottom bunk and live out of their suitcases for the six days. There would be little privacy for those in steerage. The beds had one sheet, a blanket, and a pillow with a pillowcase. The linens appeared clean.

There were two bathrooms at the end of the hall, one for men and the other for women. The bathrooms had a sink with a mirror above it and a toilet. There was one more door, the shower room. Everybody, girls and guys, had to book their time and stick to it. Gino told the boys that fights over the shower were the only times that anybody ever raised their voice.

Jack asked Gino about food. Gino told them that every morning they put out fruit in the communal area on level 1. At lunch, they put out sandwiches. Same for dinner. Gino pointed out that in steerage, only *light snacks* was written right on the ticket.

The ship started to fill. Steerage passengers could walk across the main deck all the way to the bow, but the second and third levels were roped off. The "first-class" and "second-class" passengers were kept separated from steerage people. Occasionally, Jack caught a glimpse of how the other passengers lived. They had lawn chairs to lounge around on and take the sun. Steerage passengers had to find a ledge of the ship to sit on. There were no lounge chairs. They had hot meals served to them, while steerage only smelled, with watering mouths, what the first- and second-class passengers ate.

Just before lunch, everyone in steerage was summoned to the common area on the first-level deck. This was the area where steerage people ate and socialized. It turned out that there were ten guys and five girls in steerage when the ship left. When everyone was there, the captain of the ship got up and welcomed all the steerage passengers. He went over a few rules of the ship about not causing a disturbance and such and wished everyone a good voyage.

If the sun was shining, everyone was on the deck, tanning or just hanging out. There were lots of shady ledges if you wanted to get out of the sun. If the weather was bad, everybody hung out in the common area inside, the only place to go unless you wanted to sit on your bed in the dormitory. Suffice it to say nobody stayed in their room, even on rainy days. Everyone went to the common area. Some people played cards. They played chess, and there was always a backgammon game going on. And a lot of the time, people just sat around and shot the shit. It was fun to get together with the gang every day.

Out of the five girls, Tina and Adele were from Montreal. They had been friends for years and decided they wanted to go to Israel. This was definitely the most economical way to get there. Christina and Susi were from France. They didn't speak English very well and kept to themselves quite a bit. Jack and the other Montrealers weren't fluent in French but tried to include them when possible. Then there was Kris. She was from Germany. She was traveling by herself. She couldn't speak English very well but always got her point across. She fell madly in love with Jack on first sight and let it be known to Jack and everyone else.

Usually, the rooms were vacant during the day, so Jack and Kris spent a lot of time belowdecks. By the third day, it was one rendezvous a day, if that. Everywhere Jack went—and there weren't many places to go—Kris was there.

The guys all got along pretty well. There were two guys from the Netherlands and two guys from Switzerland. They hung out together. There were two guys from Germany, who spoke perfect English, and two guys from the United States. Those were the guys Jack and Joe hung out with mostly.

After about four days on board, Jack found himself in a heavy conversation with Adele, one of the girls from Montreal. She told Jack a lot of personal things as they talked, and before they knew it, they were belowdecks in the women's dorm. Right in the middle of things, Tina walked in. She froze when she saw them, then turned and left. Adele jumped up and dressed quickly and returned to the common area. That was the end of Jack and Adele. Jack didn't understand why but was never given an explanation.

By the time the ship docked in Haifa, every one of them was more than ready to disembark. Once they cleared Israeli customs, Jack and Joe sat down at a nearby café and were joined by several of the other "steers." They began referring to themselves as "steers" by the end of their voyage. Plans and destinations were discussed for the next hour. Goodbyes were made, and everyone took off in different directions. Kris made Jack promise to meet up with her at her relative's house in Nazareth. Jack took the name and address of her cousin and said he would try but couldn't promise anything.

The boys planned to head up north to a *kibbutz* called K'far Blum. Joe's cousin lived there, and Joe was sure she would put them up for a couple of days. It was just after lunch, and the boys decided to leave Haifa and travel to Jerusalem to see the Old City. Then they could continue north to the *kibbutz*. They found out that the best way to get to Jerusalem was by *sherut*. A *sherut* was a shared taxi, usually a minivan or a large car, that could take eight to ten people at a time. You just flagged it down, and if the driver had room for you, he or she would stop. The driver would yell out, "Jerusalem!" If that was where you were headed, you jumped in. He would give you a price equivalent to the cost of a bus ride, because others were also paying.

The ride to Jerusalem was interesting. People jumping into the taxi and some jumping out. After about two hours, they arrived in Jerusalem and the driver let them off at Damascus Gate, the main entrance gate to the Old City, just as they had requested. The walls of the Old City had been built by the Turkish sultan Suleiman the Magnificent in the sixteenth century. There were eight gates into the city, seven open and one closed.

The Old City was amazing, with its different customs, culture, food, and language. The *shuk*, or marketplace, was bustling with locals buying fruits and vegetables, and tourists trying to get a bargain on authentic Arabic clothing and keepsakes. Bargaining over the price was commonplace in the Old City. If you didn't haggle over the price, you would pay a much higher price than you should have paid. The vendor expected you to bargain over the price, so he started high. In the winter months, when there were fewer tourists, you could walk out of the shop during the haggling and the vendor would call you back in and give you the lower price. He needed the sale, so it was prudent to take a little less. In the summer, when there were many tourists, if you tried to use the "walkout" tactic, the vendor would let you go. They knew the next tourist would likely pay the inflated price.

As Jack and Joe walked through the ten-foot-wide alleyways with vendors on both sides and lots of people walking in both directions, they saw a sign that said, "Hotel." They needed a place to sleep, so they went in. There was a small vestibule and a very steep staircase to the second floor. They figured the first floor was probably accessed from another door. At the top of the stairs was a large room with marble floors. There was a sofa along one wall and a large wooden desk facing the stairs. An Arabic man sitting behind the desk stood to greet them. Two ceiling fans wobbled on their highest settings, it seemed, trying to circulate the air while creating a constant hum with a tick.

The man behind the desk introduced himself as Fareed, the owner. Fareed welcomed them and asked how he could help them. They said they needed a room. Jack asked how much it cost. "One American dollar per room, per night," answered Fareed. Joe asked if one room was big enough for two people. Fareed said no without any hesitation. Jack asked if they could see the room. Fareed said, "Most welcome."

Fareed ushered them through a door off the front office into a hallway. The hallway had no ceiling or roof; it was open to the elements. On one side of the hallway were six doors. At the end of the hallway was a sink with a mirror. Fareed opened one of the bedroom doors to show them what to expect. There was a bed which took up the entire room other than about two feet beside the bed and about three feet when you walked into the room. The toilet was in another location. No wonder it was only a buck a night.

Jack looked at Joe with reluctance, but since they needed a place to sleep, they took two rooms. After Fareed left to get the keys for room numbers 5 and 6, Jack commented to Joe that they probably didn't change the linens after each guest left. Joe grunted, thinking he was probably right.

Fareed returned and gave the boys their keys. They each gave him a buck, and Fareed returned to his desk. They both wondered how Fareed could possibly make a living renting out these rooms. Only longhairs or young travelers would ever consider renting a room in such a dump. Fareed must be involved in something else, Jack thought. Maybe something on the first floor was going on that brought in some money.

Jack took room number 5, and Joe took room number 6. They dropped their bags in their rooms, took a hard look at their dreadful accommodations, and couldn't help but smile. They left the hotel and were already in the heart of the Old City. Jack enjoyed being able to walk out of the hotel and right into the marketplace. That was probably the only good thing that could be said about their hotel. They walked through the alleyways, checking out each of the shops along the way. Jack saw a shirt he liked. He decided to go in and bargain for it. The price was ten American dollars. Jack knew that was way too high. If he could bargain the vendor down to six, he would buy it. Jack offered the vendor four dollars. The vendor shook his head from side to side and said, "Eight dollars." Jack walked out. The vendor called him back and said, "Okay, seven dollars." Jack said, "Six." They shook hands, and Jack got his shirt for six dollars.

The vendor didn't take any offence. Matter of fact, with all the back-and-forth while bargaining, everyone was joking around and laughing and getting along well. The vendor's name was Mohammed. He was really friendly once the negotiations were complete. He offered them tea, which they accepted, and sat down and talked about all kinds of things as they sipped their tea. A couple of tourists came into the store, and Mohammed got up to serve them. After he made the sale, he sat down with the boys again. The last customer didn't know they could bargain with Mohammed, so Mohammed scored by getting the full price. Neither Jack nor Joe could remember how the topic of hashish came up. They might have said something, or maybe it was Mohammed who brought up the subject. Mohammed told them he could get them whatever they wanted. The boys never purchased more than a finger of hash at a time. A finger was exactly what it sounded like, a piece of hash the length and width of a normal index finger and about a quarter-inch thick. That was how it was sold on the street. One had to be very careful when purchasing on the street because of the many informers. Just for kicks, Joe asked Mohammed how much it would be for a kilo of hash. Mohammed answered, "Seventy-five American dollars." Back in Montreal, an ounce was going for seventy-five dollars or more. They could keep some and sell at least thirty ounces. They could make a killing. All they had to do was get the hash home to Canada.

The Hash Den

Without a plan in place, with no thoughts as to the consequences, they both decided, on the spot, to buy the kilo of hash. It was the thrill of the adventure. They were young and foolish, without any fear of getting caught. How they intended to travel with the hash once they got it wasn't even considered or even discussed yet.

Mohammed told them to wait while he got a friend to watch his shop for an hour or so. He was going to show the boys where they would meet the following day and receive the hash. He got his buddy to cover for him, and all three of them, under Mohammed's lead, headed out. They went to East Jerusalem in an area that had dirt roads with donkeys pulling carts and definitely not what the average tourist would see. They arrived at a house with a front door that opened right onto the dirt road. Mohammed approached the door and knocked three times. The boys watched. Seconds later, the door opened. Mohammed motioned for the boys to enter. He introduced them to his buddy, Sam, which was a nickname, because even other Arabs couldn't pronounce his real name properly.

They all sat down and discussed what would happen the next morning. At 11:00 a.m., Jack and Joe were to stand across from Mohammed's shop. They were to talk nonchalantly but keep an eye out for Sam. When they saw him, the boys were to follow him, but not too closely. He would lead them to the house they were in, and they could pay and get the hash. Mohammed asked where the boys planned to go from there once they had the hash. Jack answered truthfully, "The bus station." Mohammed said he would arrange for transportation to get them there. They thanked him, thinking that was a nice added touch. Mohammed and the boys left to go back to Mohammed's store. They told Sam they would see him at eleven the following morning. They never even asked how big a package it would be. They were caught up in the cloak-and-dagger adventure, with no regard to the details. Or consequences, for that matter.

Once Mohammed got the boys back to his store, they were familiar with the area and had no problem returning to their hotel. They thanked Mohammed and told him they would see him in the morning.

When they returned to the hotel, they were surprised to discover that three Australians had checked into their fleabag hotel. They were longhairs too, traveling around the Middle East and heading to Afghanistan. They were in rooms 2, 3, and 4. There was only one room left at the inn. The Australians had a joint rolled and offered to share it. So they all piled into room 3 and smoked the joint. Everyone started telling their stories of impressive things they had seen or done. Then, Bobby, one of the Aussies, told the boys that a few hours earlier, they had smoked hash in an Arab hash den. Jack asked where it was, thinking he and Joe would like to check it out later on. Bobby suggested all five of them go back for round 2 right then and there. There were no objections, except from Jack.

Jack wanted to see Yad Vashem before he went to the hash den, a memorial to the Holocaust. Jack was sure that if they went to the den first, they would probably never get to Yad Vashem. Bobby asked if he and his buddies could come along with them and then go to the hash den afterwards. Everyone agreed.

Yad Vashem was about thirty minutes by bus from Damascus Gate. When they arrived, they began walking through the exhibits. There were pictures of everything that happened. There was a picture of two smiling Nazis with two Jews hanging from a lamppost in the background. They showed the trains that brought the victims to their deaths. There were pictures of how they slept at night on bunk beds five levels high, leaving less than two feet to crawl into each bed. There were pictures of a room full of luggage that they took from the victims as they arrived. There were rooms of jewelry confiscated. They even had photographs of piles of gold fillings taken out of the teeth of the victims before they were executed. And there were many, many more atrocities shown in pictures and detailed by those that survived. The soldiers liberating the camps were ordered to take pictures and recorded everything, knowing that one day some people would deny the Holocaust ever took place.

When everyone made it to the exit, no one was crying, but they were definitely teary-eyed. If you were human, the Holocaust memorial had to have a profound effect on you. You didn't have to be Jewish. They discussed what they saw and how they felt. Jack and Joe knew quite a bit about the

Holocaust before they saw the exhibit, but the guys from Australia knew little of what really happened and were blown away. They all had to sit outside on the entrance stairs for about twenty minutes before carrying on.

They headed out to the hash den, Bobby leading the way. He was sure he would be able to find the place again. The alleyways in Jerusalem could get very confusing, especially once the sun began to set. Bobby did get turned around a couple of times but persevered and got them to the hash den. After all, there was no sign reading, "Hash Den." Jack wondered how they found the den in the first place but never bothered asking.

Bobby went up to the door and knocked. An Arab answered the door and asked, in English, what they wanted. Bobby replied that they wanted to smoke hash. The Arab opened the door and ushered everyone in. He led them into a room about ten feet wide by fifteen feet long. There were ten seats, with a big iron pot filled with red-hot coals right in front of the chairs. On the back wall was what looked like a boarded-up window to the front walkway. Nearby the pot of coals was a water hookah.

The first seat was occupied by an Arab. The Aussies said he had been there when they were there earlier in the day. He looked really out to lunch. Jack didn't want to crowd him, so he sat on the third chair. Joe sat next to him, and the Aussies took the next three chairs.

Everyone watched the Arab prepare the hash for the hookah. As each person sat in silence, they were mesmerized by watching the hash being kneaded to some specification only the Arab knew. When he was satisfied, the hash was placed to the side. He then picked up a red-hot coal with a homemade tool and placed it in the bowl of the hookah. He added another coal. And then another. He then placed the hash that he prepared on top of the red-hot coals. He took a few big hauls on the mouthpiece of the hookah and got it burning well. Next, he passed the hose with the mouthpiece to the Arab in the first seat. The Arab took the hose and put the mouthpiece deep into his mouth and took a big haul. He blew the smoke out and took another huge haul, held it in, and passed the hose to Jack. Jack took the hose, and without wiping the slobber off, in fear of insulting the Arab, he put it in his mouth. He took a haul and blew the smoke out, just as the Arab did. He took another big drag and kept it in as he passed the hose to Joe. Joe did the same, and so did the three Aussies. After everyone had smoked, they just sat there, getting higher every second, as the hash took effect. Jack could tell by looking at the other guys' faces that they were blitzed also, feeling a little insecure and out of control. That hash was literally mind-blowing. They were so stoned.

All of a sudden, a man in a suit pushed open the boarded-up window from the outside and jumped into the room. Everyone was totally freaked out. The man in the suit spoke to the head Arab in Hebrew, and then everyone was told to leave. Nobody knew what was happening, but they were way too stoned to object to anything. As each guy was exiting, the Arab asked if they wanted to smoke again. Everyone said no, but they all meant, "No way." When they got outside, a dozen uniformed police officers were lined up along the outside wall. The guys walked by the glaring officers without anybody stopping them. Finally, they were far enough away from the den and found an all-night coffee shop. They sat down and looked at each other in disbelief. They determined that the guy who had jumped through the window was trying to bust them. But the boys had already finished smoking. The cop had to catch them in the act. That was why the Arab asked if they wanted to smoke again. And all the cops outside were there to help with the arrests. That was the only explanation possible.

The Aussies said it was a lot different when they went to the den the first time. They thought the hash was stronger the second time around. Jack commented that he thought the hash was really strong, maybe the strongest he had ever smoked. Now that they had come down a bit, they headed back to the hotel. When they got there, everyone made some sort of comment about how insane the night had been before everyone hit the hay. The Aussies were heading to Istanbul, Turkey, in the morning. Jack and Joe had a very important meeting at 11:00 a.m.

The Aussies were up first. By the time Jack and Joe surfaced, the Aussies were all packed up and ready to go to the airport. Good wishes were exchanged, and the Aussies headed out. Jack and Joe weren't far behind. They packed up all their stuff and checked out also. They wished Fareed much happiness and good health and then left to meet Sam.

The Buy

They had an hour until they had to be at Mohammed's store. They planned on stopping at the first place they saw where they could get a coffee and a croissant for breakfast. Then they planned to continue on to rendezvous with Sam.

At 11:00 a.m., they were standing across from Mohammed's store, as planned. They pretended they were checking out the shops, always keeping an eye out for Sam. Jack saw Sam first, tapped Joe's arm, and they began to follow Sam at a distance. Once they were out of the marketplace and on the dirt road, Sam waited for them to catch up, and they all proceeded together. When they got there, Sam knocked three times. Jack thought three knocks was their obvious secret code to open the door. Mohammed opened the door, and they went in. Things weren't as calm as they were the day before. Mohammed seemed nervous. And so did Sam. Jack thought maybe it was normal for Mohammed to be uptight. Usually, he probably dealt in fingers of hash, so this was a big deal for him. How many people would buy a kilo of hash unless they lived locally and sold it? No traveller would ever buy that quantity if they intended on crossing a border.

Jack paid Mohammed seventy-five American dollars. Ironically, the Canadian dollar was worth 10 percent more than the American dollar at that time, but the world loved American currency. All the many black markets gave better deals if you paid in American dollars.

Mohammed handed Jack five flat sacks of hash. Each sack was eight inches long, five inches wide, and three-quarters of an inch thick. Jack was so naive that he stepped back and snapped a picture of Mohammed and Sam, right there when the deal was being done. Mohammed objected just as Jack took a second picture, this time capturing Mohammed trying to hide his face with his hands. Mohammed ordered Jack, quite firmly, to get rid of the camera. He motioned to grab the camera, but Jack pulled it away out of reach. Immediately after the exchange was made, Mohammed rushed Jack and Joe out the door. Things started to move very quickly, at an uneasy pace. Right outside the house was a limousine. The villagers saw the limo and thought a movie star must be visiting someone in their humble community. They began waving and cheering. Jack and Joe were so rushed they got into the limo, carrying the sacks of hash in their hands, while the villagers looked on. Mohammed instructed the driver to drop them off at the bus terminal. And off they went.

While they were on their way to the bus terminal, they decided to put the hash in their duffel bags temporarily and worry about what to do with it later. It wasn't very long before they arrived at the bus terminal. The driver said there was no cost for the ride; Mohammed had the fare covered. They said thanks and got out of the limo. As soon as the boys closed the car door, the limo took off like a jet. Jack remembered thinking that the cabdriver was in one hell of a hurry.

Jack and Joe stood there trying to figure out what to do next. Suddenly, a guy in a suit walked up to them and told them to come with him. He had eight uniformed cops standing behind him. He spoke to them in Hebrew. Jack said, “English, no understand,” even though he understood perfectly. The cop raised a finger, as if to say, “One minute.” He turned and disappeared, taking all the other cops with him.

Joe saw a public washroom, so they shot over and then down this steep stairwell leading into the washroom. Jack went into one cubicle, and Joe went into the cubicle next to him. For the first time since they decided to buy the hash, they discussed what they should do with it. They determined that they had to put it on their body. They figured the first thing the cops would do was search their luggage. It was their only chance. Jack tried to put a sack of hash in his boot. It fit. He told Joe to see if he could fit a sack in his boot. He tried, and it fit in his boot also. So they put a sack in each boot. They had one sack left. Joe offered to put that one in his underwear over his “family jewels.”

They climbed the stairs outside the bathroom to find Number 1 Cop with Number 2 Cop waiting for them at the top. Number 2 Cop spoke perfect English. He told them to come with him, without a hello or anything. The boys followed, as instructed. They entered the terminal building and went up to the second floor of the two-story building and down a long hall with offices on both sides. Number 2 Cop took them into a large sparsely furnished room. It had a long table along one wall. There was a large picture window facing the front of the terminal, with a desk and three chairs in front of it. In the corner was a small counter with a kettle. That was it.

The boys were told to put their belongings on the long table. As soon as the luggage hit the table, three or four of the uniformed cops began picking through everything. Jack thought to himself that at least they were smart enough to have the hash on their bodies, or the jig would have been up at that point. Jack was told to stand about six feet from the desk. Joe was told to sit in one of the chairs. Number 2 Cop told Jack to take his shirt off. Jack did as he was told. Number 2 Cop looked at Joe with piercing eyes while Jack took off his shirt. There was no reaction from Joe. Then, Number 2 Cop told Jack to take his jeans down. Jack undid his jeans slowly and lowered them to the top of his boots, where they bunched up, still hiding the sacks of hash from view. He left Jack standing there half-nude, while he quizzed Joe about who his Arab friends were. Joe said they had no Arab friends. He asked if they bought drugs. Joe said, "Of course not." The cops thought the boys bought that quantity of hash because they had some Arab or Israeli to take it off their hands and make a profit. It was inconceivable that anyone intended on traveling with all that hash.

Number 2 Cop told Jack to get dressed. Boy, was Jack relieved! He then told Joe to change places with Jack. Jack got dressed and sat down. Joe stood where Jack had previously been standing. The cop told Joe to take off his shirt. Joe did. Jack and Joe held their breath, waiting for Joe to be told to lower his jeans. In seconds, it would all be over. Number 2 Cop was visibly pissed off. His hands were shaking uncontrollably. He asked Jack the same questions he had asked Joe, and Jack gave him the same answers as Joe did. Number 2 Cop got up abruptly and stormed out of the room, leaving Joe standing there without his shirt on. The boys were stunned. In the moments before the cop had left the room, Jack and Joe had been waiting for the axe to surely fall. They were on the precipice of losing everything. And all of a sudden, when the bust was just about to happen, Number 2 Cop stood up and walked out of the room. The boys heard a lot of yelling in the hall as they waited to learn their fate.

While the head cop was out in the hall, Jack looked out the picture window. He had a bird's-eye view of where the limo had dropped them off. He saw uniformed policemen going up and down the stairs to the public washroom where he and Joe had been. Two uniformed cops were in a garbage bin, sifting through the garbage. Then, it occurred to him. These cops weren't just hassling them; they knew the boys had the hash. And the only way they could have known that was if Mohammed had told them. They had been caught in a scam. The informer—in this case, Mohammed—sold some hash to a tourist—in this case, Jack and Joe. Then, Mohammed notified his police buddies and set up the bust. Mohammed got the hash back for the next unsuspecting tourist, the cop got the bust, and they split the bread that the boys had laid out. Beautiful. It probably worked every time. Mohammed even arranged transportation to the "wolves' den," which Jack and Joe thought, at first, was such a nice added touch. The ride to the bus station was crucial in Mohammed's devious plan.

Jack looked at Joe standing there without his shirt. When the head cop returned, he would surely tell Joe to take down his pants, the same thing as he had asked Jack to do. When he did, it was game over. The boys knew life as they knew it would be over. It was a Middle Eastern jail for them with all kinds of prisoners that weren't crazy about Jews. They were preparing themselves for the inevitable while still trying to keep the faith.

A short time later, Number 2 Cop returned. The boys waited to see what would happen next but showed no fear even though they were scared shitless. Now that the consequences for their actions were front and center, Jack and Joe struggled with what they knew was about to happen. Number 2 Cop sat down in front of Jack and just stared at him with those piercing eyes of his. He totally ignored Joe. With a snarl, he told Jack to pack up his belongings. Jack got up and walked over to the table and began packing up his things, watching closely what was happening to Joe. Number 2 Cop stood and walked over to Joe. Instead of telling Joe to lower his jeans, as suspected, he told him to get dressed and pack up his stuff. Then, he stormed out of the room again. Joe put on his shirt and joined Jack at the table where their luggage had been strewn about. All the uniformed cops stood around, watching the boys pack up. Jack's and Joe's hearts were pumping so hard they thought the cops could hear them beating. Number 2 Cop knew Joe couldn't have five sacks of hash in his underwear, so he assumed the hash wasn't on their bodies. What a blunder!

When they were all packed up, they were told they could leave. The boys felt like they were in the twilight zone. When they exited the building, there were cops everywhere. Jack went to the ticket counter and purchased two tickets to Tiberias. Tiberias was on the way to Kiryat Shmona. Jack didn't want Number 2 Cop to know where they were really going. When they got to Tiberias, they could transfer to another bus to Kiryat Shmona.

They went to the bus stop where the bus to Tiberias pulled in. It was the first time the boys were able to talk to each other since they had been detained. They talked about the miracle they had just experienced, a miracle in the holy land. They both chuckled. Jack told Joe he figured out they were scammed by Mohammed. Joe had already figured it out too. It was probably the first time ever that their little scam hadn't worked. It felt good to fuck them for a change, on behalf of all those in jail, doing hard time because of their poor judgement. The price for getting caught was super high.

Number 2 Cop couldn't leave it alone. He was obsessed with figuring out where Jack and Joe had hidden the hash, and that made him furious. While Jack and Joe waited at their bus stop, Number 2 Cop, with his entourage, approached them again to hassle them further. He asked Jack what he had around his neck. It was a beautiful brass roach clip, but Jack lied and said it was a family keepsake. As Jack was answering the question, the head cop took ahold of the roach clip and pulled up Jack's shirt at the same time. If he had done that to Joe, he would have seen the top of the sack in Joe's underwear.

The bus finally arrived. The boys couldn't wait to get out of there. The cops had to let them board the bus. Once they were seated, they looked out the window and saw

Number 2 Cop staring at them, totally baffled about the whereabouts of the hash. Jack felt like waving but didn't, of course. Why kick an angry bear in the nuts?

The Kibbutz

The bus ride was fascinating. The radio was blasting all the way. All the Egged buses were like that. Every bus was owned by the Egged Bus Company and was always referred to as an Egged bus. You heard the odd song, like Creedence Clearwater Revival or Cat Stevens, or "Rolling on the River" by Tina Turner, which were very popular in Israel at that time. But everybody was listening to the news. After all, an attack could happen at any moment in Israel. Israel didn't only have to worry about being attacked by a neighbouring country; there were also small groups of terrorists who jumped across the border to cause as much turmoil as possible and kill as many Israelis as they were able.

People from every walk of life got onto the bus. An old lady got on the bus, carrying a sack with at least two live chickens in it. Jack had no idea if she was going to sell them or if she had just bought them. She got off at the next stop. She had probably gone from her village to a neighbouring village to get the chickens. At some point, a soldier returning home from training got on the bus. He sat next to the boys. He was carrying an Uzi machine gun, as every other soldier did. Just before he got off the bus, he asked Jack to hold his Uzi while he tied his shoelace. Jack did. Jack waited until he was done and returned the Uzi as if it was no big deal. But to Jack, he was holding a machine gun and it was a really big deal. When the boys were in Germany a few weeks earlier, Jack was talking to a longhair who had been to Israel. He told Jack that there were soldiers everywhere carrying machine guns. In Jack's mind, he had pictured soldiers goose-stepping down the street, but it wasn't like that at all. Soldiers were hitchhiking everywhere, and they all carried a machine gun slung over their shoulder. The Israelis picked them up constantly. It was their national duty. Everyone felt safe. Even though they could be attacked at any moment, life carried on and there was more of an atmosphere of safety than of fear.

Israel was a lush, man-made paradise which had once been mostly desert. It was an ancient land with churches, synagogues, and mosques dating back hundreds, if not thousands, of years. Israel has religious significance to Jews, Muslims, and Christians. It was difficult to travel anywhere in Israel without seeing something historic. There were sights to see straight out of the Bible.

The trip to Tiberias took about two hours. They got off and transferred to another bus going to Kiryat Shmona. Tiberias was a beautiful resort city on the Mediterranean Sea. There were sandy beaches if they wanted to go swimming, but they chose not to have to deal with the hash. There were lots of ancient sites to see, but the boys just got on their bus and headed north.

Kiryat Shmona is in northern Israel, between the Syrian and Lebanese borders, in the region of Upper Galilee. Kibbutz Kfar Blum was in the Hula Valley, about six kilometres southeast of Kiryat Shmona. Once they arrived in

Kiryat Shmona, they would have to take another bus right to the kibbutz.

It was their first day traveling with the hash in their boots. They couldn't wait to take their boots off. They both felt like they had blisters on their ankles from the constant rubbing as they walked. The weather was really hot, which added to their discomfort. They couldn't just go home and change. Everyone was wearing sandals, and they were wearing boots.

As soon as they got off the bus, a *kibbutznik*, a person who lives on the kibbutz, greeted them and asked if he could assist them. Apparently, it was common for someone to be watching the bus stop, just to see who was coming into the kibbutz, hoping to weed out undesirables. Joe told him they were there to see his cousin Ellie Weiss. The kibbutznik knew her and offered to walk them right to her house. They now had status, being a relative of another *kibbutznik*. The offer was very much appreciated. Jack wondered how Ellie would react when they showed up on her doorstep unannounced. Surprise!

The kibbutz was huge, especially if you didn't know where you were going. It was around suppertime, and Ellie was just on her way to the dining hall when they arrived at her door. When she first saw Joe, she was shocked. Once that registered in her brain, she was honestly happy to see Joe and invited them in. She thanked the guy who had brought them there, and so did the boys.

Once they got in and were seated, Joe introduced Jack to Ellie. Joe apologized for showing up without calling. They had grown up most of their lives together, so they had a close relationship. Ellie said she was glad that Joe had come.

Ellie wanted to hear their plans and where they had been, but first, they had to go for supper before the suppertime was over. On the kibbutz, you could cook and eat in your house if you wanted to, but most people ate in the communal dining hall. Everything was prepared for you. You ate anything and as much as you wanted and then left. It was communal living, with everyone doing their job, and the wheel kept on turning.

Ellie heard through family members that Joe was going to Israel and might look her up, but she never really expected him to show up. Not only did Joe show up, but he also had a buddy with him. But it was all good. Ellie got on the case and talked to the right person, who found them a place to stay. They got a room, just for the two of them, in the volunteers' accommodations. Volunteers came from all over the world to help harvest the crops or do whatever job was assigned to them. Their workweek was six days. If you worked in the fields, you got off on Saturday, the day of rest. If you worked in the dining hall, for example, you might work on Saturday and take off another day in lieu. As payment, workers were fed, clothed with any necessary clothes for their job, given two packs of Gitane (black tobacco cigarettes) per week and fifteen lire, which was about five bucks. You were able to have a steak in Kiryat Shmona for less than fifteen lire.

Ellie was proud of herself for putting everything together for the boys. She suggested that they all go have dinner together in the dining room, and then the boys could go to their room and get settled. There would be lots of time to talk. The boys agreed, even though they couldn't wait to get back to their room and deal with their feet and, of course, the hash. Ellie and Joe reminisced about old times. Ellie talked about her new life in Israel and what it was like living there. Joe told her about their travels and all the places they had seen. Jack did a lot of listening.

Soon, the day had taken its toll, and Jack and Joe excused themselves. They made a date to meet in the dining room for breakfast the next morning. Ellie offered to walk them to their room on her way back to her house. The boys graciously accepted. The sooner they got there, the better they liked it. As soon as Ellie got them to their room and left, the boots came flying off, followed by the hash. The hash had slowly crept down and rubbed both their ankles. This could develop into a big problem. It was only their first day traveling with the hash. If they started limping, it might attract someone in customs. An agent could easily ask them to remove their boots. If that ever happened, it was lights out.

It was a relief to be bootless. They knew that they had to devise something that would fix the ankle problem, but they had time to think about it. First, they had to deal with the hash. They decided to crash and worry about it in the morning, after they met up with Ellie for breakfast. Jack asked Joe how long he thought they were allowed to stay on the kibbutz. Joe said he would ask Ellie. The next morning at eight o'clock, Jack and Joe met Ellie. They started breakfast off with a grapefruit. It was the best grapefruit Jack had ever eaten. All the fruit in Israel was grown right there, so everything was scrumptious.

Fried eggs came around, and everybody took two. There was a choice of sunny-side up, over easy, or well done. Jack saw some young *kibbutzniks* taking five or six eggs. They burned it off working in the fields. If you wanted toast, there were a number of toasters on a table against the wall, and you had to get up and make your own. There were three or four different breads and an assortment of rolls.

Joe took toast orders and offered to make everyone's toast for them. While he did that, fried potatoes came around, and they all loaded up, Ellie taking some for Joe and putting them on his plate. They all got coffee, and by the time Joe finished making the toast, they were ready to dig in.

After breakfast, Ellie went to work. She worked in the day care. As she turned to leave, Joe remembered to ask her how long they could stay before overstaying their welcome. She said at least a few days, then hurried off. Guests were usually required to stay at the guest house and had to pay for each day they stayed, which included their meals in the dining room. Jack and Joe were sort of personal guests of Ellie, a *kibbutznik*. So the rules were bent, and the boys were allowed to stay without working. And they didn't have to pay anything either.

Jack and Joe set out exploring. They got lost in no time, but they didn't care. There weren't very many people around at that time of the day. Many of the jobs started at 4:00 or 5:00 a.m. and were finished by 11:00 a.m. Then, volunteers could do whatever they wanted until the next morning at four. Apparently, there was an Olympic-size swimming pool that many volunteers and *kibbutzniks* took advantage of.

As they walked around aimlessly, they bumped into a couple of girls from the United States. They were twins, even though they didn't look anything alike. Their names were Amy and Barb. They were both very attractive in their own way. They had come to Kfar Blum as volunteers about a month earlier and had planned on staying until they had to go back to university in September. They worked in the dining room and started work at noon and got off at six, so they had lots of time to get acquainted. They walked around the kibbutz for a while, with the girls pointing out things they thought the boys should know, like where the dining room was, the location of the pool, Ellie's house, their room. The boys actually had a good idea of the layout of the kibbutz after that tour.

Then, the girls led Jack and Joe down a narrow path to a river running alongside the kibbutz. It was the river Jordan. the same river Jordan from the Old Testament. The water ran down from Mount Hermon and flowed south along the border between Syria and Lebanon, and through northern Israel to the Sea of Galilee. Because it ran downhill, when it passed the kibbutz, the current was quite strong. Amy and Barb said it was fun to walk as far north up the river as they could, then jump in and let the current carry them back down to the kibbutz.

The girls had to go to work. They suggested they could all have dinner together, if the boys wanted. The boys definitely wanted to. They arranged to meet at the dining room at six.

The boys decided to go back to the room, get the hash, go back to the river Jordan, and bury it there until they left the kibbutz. They could break off a piece to smoke and bury the rest.

They returned to their room and put the hash in an opaque plastic bag. They took a leisurely walk back to the river Jordan, holding the plastic bag as if there wasn't anything significant in it. And of course, they bumped into Ellie with a dozen or so little children from the day care. Jack and Joe greeted Ellie. One curious little girl asked Jack what was in the bag he was carrying. Jack felt his face flush red and hoped Ellie didn't pick up on it. Jack answered the little girl, telling her he was returning some clothing he had borrowed from a friend. The little girl asked Jack if she could see it. Jack told her that was impossible because it was all wrapped up. The little girl accepted Jack's explanation, and the conversation ended there.

They took a few wrong turns but made it down to the river. No one was anywhere in sight. They saw two large rocks with a space between them—a perfect place to bury the hash. Joe kept watch, while Jack began to dig. He used a stick at first, then a rock, and then his bare hands to dig out an area one foot by one foot by one foot deep. Jack placed the hash, still wrapped in the plastic bag, in the hole. He pushed the dirt back into the hole. He searched the area for some branches and covered up everything so nobody would be the wiser.

Joe sounded the alarm. Someone was coming. Thankfully, the burial was finished, and the branches completely camouflaged the area. The intruders were a group of volunteers. They also worked in the dining room and had the day off because they had worked the previous Saturday. They explained they were just hanging out by the river until lunch and that, after lunch, they were going to spend the afternoon by the pool. They said if the boys were up to it, they could all have lunch together and then go to the pool with them. Jack and Joe felt like they were on a resort.

The volunteers had their bathing suits on under their clothes, but the boys had to go back to their room and change. After lunch, the volunteers headed to the pool, and the boys went to their room, changed, and met up with them soon after.

The pool was beautiful. It had a high diving board, which allowed some of the *kibbutzniks* to become pretty good divers. It was entertaining to watch their impressive dives as they showed off for the ladies.

Mount Hermon stood high in the background with snow on its peak, even though it was hot in the valley where they were. It was the highest point in Israel and could be seen from most of northern Israel. The mature trees and gardens surrounding the pool had been somebody's vision years before. Now that vision with Mount Hermon's peak covered in snow standing tall in the background could easily be a postcard. The boys spent the entire afternoon swimming and taking the sun. The volunteers they had lunch with introduced them to at least a dozen other volunteers who came to lounge around the pool after finishing work. They hailed from Canada, the US, Chile, Pakistan, Australia, France, Germany, and lots of other countries.

It was almost suppertime. They said goodbye to all the friends they had made and headed back to their room to shower and change. They had to meet the twins for supper at six.

They got to the dining hall and saw the girls waiting for them by the entrance. When the girls saw Jack and Joe, they came right over and hugged them. Jack liked Amy, and Joe liked Barb. The menu was spaghetti and meatballs. They made a huge salad they could all share to go with their spaghetti. Jack walked over to a side table and took four dinner rolls and brought them back to the table.

When dinner was over, they all went back to the girls' room. The girls had their own room; there were no other roommates. Jack and Joe never made it back to their room until the girls left for work the next day. They made a date to meet for supper again and were off. The boys were up most of the night and needed to have a little shut-eye before venturing out again. Too bad the girls had to go to work. They must have been as tired as the freeloading boys were.

Life was grand on the kibbutz for Jack and Joe. But they began feeling like they were starting to take advantage, not pulling their weight like everyone else. They didn't want to put Joe's cousin Ellie in an awkward situation either. So they made the hard decision to leave the next morning and return to Jerusalem. There were so many things they still wanted to see in the Old City and surrounding areas.

First, they had to find Ellie and let her know their plans and thank her for all her help. She would probably be relieved that it never came to asking them to pack up and leave. Then, they had to dig up the hash so they would have it when they got dressed the next morning. They discovered, after much thought, that a sanitary pad placed in their sock beside their ankle would solve the chafing problem. They had to say goodbye to Amy and Barb. The boys were not looking forward to that. They could get all their chores done and then tell the girls that they were leaving when they met them for supper. Once they found Ellie and told her what their plans were, they intended to spend the rest of the day at the pool. But first, they had to get the hash. If people were down by the river, they could always go back later. They got there with no problem. They seldom got lost anymore. Too bad they had to leave!

Joe stood guard again. Jack removed the branches and began digging the loosened dirt with his hands, like a dog digging a hole. He was praying nobody would come for a few more minutes. In no time, Jack reached the bag of hash, pulled it free, and filled the hole with dirt. Carrying the hash in the bag nonchalantly, they returned to their room. Once there, they put all five sacks into the bottom of Jack's duffel bag.

It was time to find Ellie. She was playing with the children outside the day care. The kids she was watching were around six years old, so she was able to talk to the boys, always keeping an eye on the children. Joe told Ellie they decided to leave the next day. Ellie told Joe how fantastic it was to see him. Ellie told Jack it was great to meet him. The boys both thanked Ellie sincerely for everything she did to make their stay at Kfar Blum memorable. They all hugged. The boys could see the relief in Ellie's face. She knew Jack and Joe thought they were living on a resort with young people from all over the world to party with. She was right—they did. But they appreciated those good times and made the hard decision to leave before they overstayed their welcome.

The boys were free to hit the pool, all chores complete. There were a lot of different volunteers there. They met a lot of lovely ladies and a bunch of cool dudes.

Before they knew it, they had an hour to get back to the room, shower, and dress so they could meet Amy and Barb for supper. They arrived a couple of minutes early. At 6:00 p.m. on the dot, the girls showed up, straight from work. They looked tired but fantastic at the same time. They went into the dining hall and picked a table to sit and have their supper. They all ate the chicken and rice with a salad. Between volunteer after volunteer stopping by their table to say hello and the girls each recounting what had happened that day at work, Jack and Joe didn't say anything about their plans to leave. They finished dinner and all went back to the girls' room for the night. The girls were totally wasted and needed sleep.

As soon as they were all in the room, Jack looked at Barb and Amy and said, "We are leaving in the morning."

Amy and Barb knew the boys were going to be leaving one day soon. It was really sad, but somewhat expected when they were told. They exchanged addresses and made promises to stay in touch. They spent an awesome night together and all went for breakfast in the morning. Jack insisted on saying their goodbyes there, in the dining hall, instead of at the bus stop. That way, the boys were able to go back to their room alone, without the girls. Jack didn't want to have to deal with the girls when they loaded up their boots. There were tears and kissing and hugging. Everyone knew deep down that they would probably never see one another again.

The boys returned to their room. They loaded up their boots, putting the pads over each ankle inside their socks. They were ready. The bus for Kiryat Shmona was arriving in another twenty minutes.

The Old City

The bus was right on time. They boarded. It was only six kilometres to Kiryat Shmona, but the bus made several stops. Along the way, the boys were able to see some other settlement communities, called *moshavs*. The difference between a *moshav* and a *kibbutz* was that on a *moshav*, the community was made up of individual farms. The people on a *moshav* had a relatively large degree of economic autonomy and often pooled their crops for mutual assistance. The kibbutz was like one big farm with the people living in a communal environment.

The trip took forty-five minutes. Kiryat Shmona was established in 1949, just after Israel got its independence in 1948. It was a small town then and, being close to the Lebanese border, had its share of hits by *katyusha* rockets and attacks from infiltrators sneaking over the border.

They had a couple of hours to kill until their bus to Jerusalem was scheduled to arrive. Jack wanted to make sure the maxi pads hadn't slipped and were still protecting his ankles, so they went looking for a washroom. They found one. Everything seemed to be in place. They went back to their bus stop, and while they waited, they discussed what they were interested in seeing in the Old City. The first thing they had to do was to go to the Wailing Wall. Joe said they could go to the Dome on the Rock. And Jack said they could go to Bethlehem. And Joe said…

They both laughed. There was so much to see.

They decided to go back to that fleabag hotel, because it was cheaper than cheap and right in the heart of the Old City. It would be a good base for them. And they could see Fareed. He seemed like a nice man. Jack couldn't understand why he sat at his desk at ten o'clock at night when he had no guests and the alleyway outside was dark. He was sitting there when they came in from the hash den the last time they were there. Jack was sure the answer was on the first floor.

Their bus pulled up. The boys waited until the entire bus emptied, and then got aboard. The bus driver announced that the bus would leave in five minutes, and then he took off. He returned with a coffee and a pastry, and within minutes, they were off to Jerusalem.

Some hours later, they arrived in Jerusalem. Instead of staying on the bus until the end of the line, which would be the main bus station, they had the option of getting off the bus near the Old City. Another passenger told the boys that was what he was doing, so the boys decided to do the same. The thought of returning to the bus terminal was unimaginable, especially as they were still carrying the hash.

The bus driver yelled out, "Old City." Jack and Joe and the other passenger all got off the bus. The boys were hungry, so they stopped to eat a falafel on the way to the hotel.

When they walked into the hotel, Fareed was so happy to see them, ecstatic to see a repeat customer. It was no surprise that no one else was there. Fareed gave them the same rooms as last time. Joe made a sportsman's bet with Jack that the linens hadn't been changed. Jack declined the bet.

They put their things in their rooms and started to look for a place to stash the hash. There weren't any vents in the room. There was no place in the room they could hide it. They had to find a hiding place, or they couldn't go out to see anything. They were constantly harassed by the police because they had long hair. Not to mention they would have to take their shoes off to visit the Dome on the Rock, an Islamic holy site on the Temple Mount. Joe suggested putting the hash under the mattress, as there was no fear of the linens being changed. But Jack thought if anybody did try to rip them off, under the mattress would be the only place anyone could look. They walked out of their rooms into the open hallway. There was no roof, so there had to be a drain to catch the water in the rainy season. The drain sat right in the middle of the floor. Jack picked up the grate quite easily over the drain and saw a five-inch ledge all around the drainpipe. The sacks of hash fit like a glove. Jack replaced the grate over the drain, and their problem was solved. They were free again.

The first place they went to was the Western Wall, a site where Jews went to pray and reflect. The Western Wall is often referred to as the Wailing Wall because of the Jews who went there to weep over the destruction of the First and Second Temples. The Second Temple was built by Herod the Great, and the Western Wall is all that remains, following its destruction by the Romans in the year 70 CE. The wall stands nineteen metres high, and every crack between the stones as high up as people can reach is jammed with messages and prayers written by the millions who have gone there to pray.

The wall is part of the Temple Mount and is only a three-minute walk from the Dome on the Rock. That was where they would go next. In the meantime, the security was high at the wall, and they had to wait in line for several minutes. When it was their turn, they walked up to the wall and touched it. Orthodox Jews were bowing and shachling as they prayed, almost as if in a frenzy of some sort. Jack and Joe just stood there, experiencing the incredible feeling of touching something ancient.

They left the Wailing Wall and walked to the Dome on the Rock. The Dome was originally built as an Islamic shrine for pilgrims, as opposed to a mosque. It was built on the site of the Second Jewish Temple. Women must cover their shoulders and must wear a long skirt or pants. Headscarves were not required. Females and males alike were not permitted to wear shorts. No Christian or Jewish religious paraphernalia were allowed. Those were the rules.

When they entered, they had to remove their boots. There were dozens of pairs of shoes on the floor by the entrance. Jack thought this was one time he was happy to be wearing boots. They were easy to spot, and no one would steal them. They didn't want to wear their sandals in case that was too informal, so they wore their boots. That was all they had: sandals or boots.

Today, Kafirs, non-Muslims, are not allowed in the Dome. Nonbelievers were allowed back then, providing they followed the rules.

Once they had seen the beautiful architecture inside and outside, well-known as an architectural masterpiece, they were exhausted and chose to leave. There was so much more to see, but you could stay there for days and not see it all. Their excursions, as tourists, were over for the day. As they headed back towards their hotel, they poked around in the shops along the way. The vendors were so friendly. It was the beginning of the tourist season, but the market wasn't that busy yet. All the merchants had time on their hands to try to coerce you into buying something.

The boys walked by a shop that sold Arabic shirts. One of these embroidered shirts caught Jack's eye, so in they went. The vendor greeted them and insisted they sit down. They did. As soon as they sat down, their host yelled some instructions to his worker, in Arabic, and two bunches of green grapes were brought out. One bunch was handed to Jack, and the other to Joe. The boys both tried one and looked at each other and smiled. The grapes were delicious. As they continued to eat the grapes, Jack began to bargain for the shirt he wanted.

The vendor's name was Abou. He kept on motioning with his hands to his mouth, telling the boys to eat more grapes. They kept on eating grapes or risk insulting him. When all the bigger and sweetest grapes in the bunch were eaten, those bunches were taken away from them, and two new bunches with larger grapes came out to replace them. Jack thought this was getting ridiculous. Jack and Joe put their hands up to say no. Jack patted his stomach and made a painful face. Abou finally relented.

Jack got up and pointed to the shirt he liked. Abou said twenty lire, but for Jack, only eighteen. Jack knew he couldn't use the "walkout" tactic, so Jack just went right down to ten lire. Abou laughed. Abou said his final price was sixteen. Joe found a shirt that he liked also, so Jack took Joe's shirt and threw it on top of the one he liked. He told Abou twenty-eight lire for both. Abou smiled and agreed.

They thanked Abou and his worker, took their shirts, and set out to find something to eat. There was a restaurant around the corner that advertised steak and lamb, amongst other things. The Arab owner was extremely friendly, sitting down at their table as he asked for their order. They weren't sure what to have. The owner, Hamid, spoke a broken English but communicated that he had a wonderful lamb dish, the most popular on his menu. The boys thought that sounded fine and ordered what he had suggested. They added a couple of Goldstar beers, one of the most popular beers in Israel, to start the meal. Hamid got up and returned with the two beers and three shots of arak, Israel's national liquor. He proudly announced that the shots of arak were on the house and went right into a toast in his broken English, "Peace, happy, and nice people." They all drank the arak down the hatch.

It wasn't long before the meal came. It was a mixture of rice and lamb and vegetables all mixed together in a heap. Jack liked to mix everything together, anyway. It was absolutely delicious. They called Hamid over and told him the meal was great; they really enjoyed it. He beamed. They ordered another couple of beers and another three arak shots.

When the araks came to the table, Jack hoisted his glass up as if to toast but waited until Hamid and Joe raised their glasses too. Jack toasted to friendship. They all clinked one another's glasses and down the hatch went another.

The boys paid their bill, thanked Hamid for a good meal and good company, and left for their hotel. There was still more they wanted to see, so they decided to stay another day or so.

They got back to their hotel, and there was Fareed, sitting behind his desk. There were still no guests except for Jack and Joe. Jack told Fareed that they had decided to stay another night, or maybe two. Fareed was delighted.

Once they got into their rooms, Jack felt like smoking a joint. He had rolled one earlier and had left it in the room. They went out into the hall, which was virtually outdoors, pulled out the joint, and smoked it. If they had heard someone coming, all they had to do was throw the joint into the drain.

Before they retired for the night, they discussed the plan for the next day. Both Jack and Joe wanted to experience a Turkish bath. Jack wanted to go to Bethlehem. First thing in the morning, they would head to Bethlehem. They could go to the Turkish bath anytime, here in the Old City.

They got up early, washed up at the sink in the hall, brushed their choppers, and left the hotel.

Bethlehem is a Palestinian town in the central West Bank, about ten kilometres south of Jerusalem. The boys were able to jump on an Egged bus and get there in a half hour. The security was very light once they got there. There were guards checking bags for bombs or weapons, but no producing passports or answering questions.

Bethlehem is the biblical birthplace of Jesus. It is a major Christian pilgrimage destination. The boys saw the Church of the Nativity. It contains a grotto, a natural cave, that holds prominent religious significant to Christians of various denominations as the birthplace of Jesus. Then, they went on to the Manger Square, which was breathtaking. It is flanked by the Church of the Nativity and the Church of St. Catherine. It didn't matter what religion you followed; it was all in the believing.

They saw a lot in Bethlehem. Joe wanted to go back to the Old City to see Via Dolorosa, often translated as the "way of suffering" or "sorrowful way." It is a processional route through the Old City of Jerusalem, believed to be the path that Jesus walked on the way to his crucifixion by the Roman soldiers.

They jumped on an Egged bus and returned to the Old City. They went straight to the Via Dolorosa. It started in the Muslim Quarter, beside Lion's Gate. Pilgrims have been tracing the steps of Jesus since before the eighth century. The walk is about one and a half kilometres. It took Jack and Joe about two hours. They didn't bother stopping at the "fourteen stations" like the really faithful did. These stations paid homage to the events that led up to what happened before the crucifixion and burial of Jesus of Nazareth. They were really proud that they got a chance to do the walk. If Jesus really had walked on the very stones that they were walking on, it was possible that they could have walked in Jesus's footstep. That was pretty cool.

The only thing they wanted to do before they left Israel was to go to the Turkish bath. They remembered they had seen a sign for a Turkish bath down one of the many pathways they had wandered through. Jack thought he could remember how to find the place. It definitely wasn't a bathhouse that the hotels sent the tourists to. This bathhouse was more for the locals.

They found the bathhouse and went in. It cost fifty cents each. They paid one American dollar. They were led into a changeroom where they could undress and put their clothes in a locker. They were each given two towels. The locker had a padlock with a key on an elastic band so you could carry it on your wrist throughout the process. You had to get totally nude. There were no women. Women had their own times to go and enjoy the baths.

Leading out of the changeroom were the showers. Jack and Joe took a nice, hot shower. That alone was a treat. They left the showers and came to a pool. No one was in it. Jack bent down to test the water, and it was cold. An Arab walked by and motioned to the boys to jump in. They did, trying to demonstrate they weren't wimps. The water was really cold. They swam a lap and jumped out. That dip in the cold water was to close up their pores, according to their newfound friend, Raazi.

From the cold pool, Raazi brought them to the next pool. Again, Jack bent down to test the water, and it was much warmer. They jumped in and relaxed for a short while. Raazi passed by and motioned with his hand to get out of that pool and get into the final pool. It was hot, like a very large hot tub. This pool was to open up their pores. The boys got in. It was bliss. They knew they shouldn't stay in too long, so after about ten minutes, they got out. Next was the sauna. There were three levels of wooden benches. Two elderly Arabs sat on the third level. Another Arab sat on the second. No one sat on the first. Jack and Joe went right up to the third and hottest level. It was way too hot for them. They couldn't believe those Arabs could stand it being so hot, without passing out. The boys got up immediately and retreated to the second level. In seconds, the second level proved too hot for them, and they moved down to the first level. That was more their speed. It was still extremely hot for them.

When they left the sauna, they were ready for bed. They felt incredibly relaxed. There was a room to just relax in before you left, but they had to shower first. They did and went back to the lounge, so to speak. They drank tea as they sat there on these comfortable benches upholstered in a soft, velvety material. There was no time limit. They could stay as long as they wanted. Not bad for fifty cents.

After they had enough, they returned to the dressing room and got dressed. They were so relaxed they yearned for a bed to lie on, even though their bed at the hotel wasn't the finest. It didn't matter. They were both happy they had decided to go to the bathhouse. The experience was wonderful.

They got back to the hotel. They saw Fareed and exchanged pleasantries before excusing themselves and going back to their rooms. They slept like babies that night.

The next morning, the boys washed up in the hallway. As Jack stood at the sink, he looked at the drain right behind him and felt sort of smug inside, thinking they had been so clever finding that stash. On the way out to breakfast, they saw Fareed again. He was sitting behind the desk, where he spent a great deal of his life. The boys informed him that they were leaving. They were going out for breakfast, and then they would return for their things before leaving for the airport. They settled up their bill with Fareed and left.

On the way to a restaurant, they passed a travel agency. They went in and booked a flight to Munich, Germany.

They were lucky to get a great last-minute deal leaving at three o'clock that afternoon, aboard a Lufthansa flight. The flight to Munich would take about four hours. It would take them about thirty-five minutes to get to Ben Gurion Airport by *sherut*, or they could take a bus going directly to the airport. They decided to take a *sherut*, once they found out that the airport bus left from the main bus terminal. The bus terminal was the last place on this earth where they wanted to be.

They had time, but they had to boogie. There was a lot to be done. They finished their breakfast and returned to the hotel. Fareed greeted them, but instead of remaining behind his desk as he always did, he came around and walked the boys back to their rooms. Fareed kept on talking to them and wouldn't leave them alone, making it impossible to retrieve the hash. He was actually standing right on the grate as he joked around with the boys.

Speaking to Joe in pig latin mixed with double Dutch, which he was positive Fareed couldn't understand, Jack told Joe to make up some kind of excuse to get Fareed to the front office. Joe walked to the front and turned around and summoned Fareed. Fareed went to the front to see what Joe wanted. Joe began to weave this ridiculous double-talk that totally confused Fareed. In the meantime, Jack was free to pull up the grate, remove the hash, and replace the grate. Jack stuffed the sacks in his duffel bag and went to the front to let Joe know it was all clear. Fareed was so mixed up by the time Jack got to the front he chose to stay behind his desk when the boys went back to their rooms. Fareed looked so confused it was hilarious.

They both went into Jack's room. Out came the hash, and the old one-sack-in-each-boot routine, with Joe putting the fifth in his underwear, as usual. They grabbed their bags, and off they went. On the way past the front desk, they stopped and said a final farewell to Fareed.

Flight to Munich

They arrived at the airport about two hours before the flight was scheduled to take off. This would be their first time crossing a border carrying the hash. As always, Jack and Joe walked confidently, showing no fear. The first hurdle was the metal detector. Jack went first. The Israeli customs agent put the wand up and down his body. When the agent waved the wand over his boots, the thing went off. Jack immediately lifted the bottom of his jeans and pointed to the zipper of his boots, praying that the agent wouldn't ask him to remove them. He didn't. He nodded in agreement and motioned Jack forward without saying a word. Joe was right behind him. The agent scanned Joe's body, and when the wand passed over Joe's boots, the same thing happened. Joe immediately showed the zipper of his boot, and the agent waved him through too. It occurred to Jack that another agent might have asked him to take his boots off.

Next was the luggage search. The agents checked every corner of their luggage, found nothing, and ushered them towards the passport agent. They approached the passport agent together. The agent wanted to know why they were in Israel. They told him they had come to see the country. He wanted to know where they were going. They said Munich. He asked them if they had bought anything. They told him some souvenirs. He stamped both passports and waved them through.

They waited at their gate for a while and finally boarded.

After about two or three hours, their feet began to swell, as did everyone's feet during a flight. But it caused Jack and Joe a lot of pain. They had to sit there and bear their discomfort without saying a word. They couldn't wait to land. The hash was starting to get soft from their body heat, and it began to smell. Jack and Joe noticed it. They were hoping no one else did.

When the plane finally landed, all the passengers gathered up their things and formed a line to exit the plane and continue on through customs. As the boys waited in line, three customs officers approached them. They directed them out of the line and told them to wait off to the side. They did as they were told. They stood there on the side and watched as every other passenger cleared customs. The situation looked bleak.

An officer wearing several medals approached Jack and Joe. He had another twenty uniformed agents standing behind him. When do you ever see something like that?! The head guy with the medals put his hand out and said, "Give me your hashish." All the other agents watched to see what would happen next. Jack couldn't speak. He was in shock. Joe blurted out that they didn't have any hashish. Jack thought that was a pretty lame thing to say. Clearly, the officer must have known the boys were carrying the hash. Then, the head guy said, "And your opiates!" with his hand out again, as if to receive the pills. Instantly, in a fraction of a glance at each other, Jack and Joe both realized that the agents were fishing. They had no idea that the boys were carrying hash. Jack got his voice back and said they had no opiates.

Jack wondered why there were so many agents standing around and came right out and asked. It turned out that all those agents were in the graduating class for custom agents for that year. Because Germany was still cold, the boys were the first longhairs coming through the border. Longhairs usually went to warm climates during the winter. They had used Jack and Joe as guinea pigs to show the agents how to deal with people like them, longhairs. After a few more questions, they had their luggage checked. Four agents searched the bags thoroughly, two searching and two observing. By the time they were getting their passports stamped, they had become friendly with several of the would-be agents, joking around and such. The boys were the only passengers left that they had to deal with. The agents were wishing Jack and Joe good luck in their travels and were giving them the thumbs-up as they exited the customs area. If they only knew…

As the boys walked out of the airport, they patted each other on the back, but only with a knowing smile. It was over. They were in Munich. The first thing they had to do was to find a cheap hostel, a place to stay until they got their bearings. Joe was running out of money and wanted to phone home and arrange to get some from his parents. In those days, there were no bank cards. The money had to be wired to a bank. One or two days later, you walked into the bank, explained why you were there, showed your passport, and left with your bread.

They found a place that rented rooms by the night for four American dollars. It had a double bed. The four bucks could be for the two of them, so they took it. The room was in a house in a nice area that had been converted into a whole bunch of rooms. Their room was number 13. It was tidy and clean, with a double bed, a dresser with a mirror, and a sink where they could wash up without leaving the room. Bathrooms and showers were shared by all the other people on a first-come, first-to-use-the-facilities basis.

The second they got to their room, they ripped off their boots and removed the hash. What a relief! The maxi pads were doing their job, so their ankles were in good shape. Now, they had to find a stash. They would probably be staying there for a few days, considering that Joe had to wait for his money to be wired to the bank. They could do the tourist thing until the money got there.

There was a vent in the wall. The screws were painted over, so they were reluctant to use it. The owner might notice that the paint had been chipped off the screws and investigate. Joe had to go to the bathroom. The bathroom was right next to their room. There were a lot of people using the bathroom, so the owners had installed a fan, vented to the outside to combat any foul odours. Joe thought the vent cover could easily be removed. Joe finished his business and returned to their room. He told Jack about the vent in the bathroom. They gathered up the hash, and both of them returned to the bathroom. Jack stood guard outside the bathroom. Joe was right. The vent cover came off easily, as expected. The five sacks of hash slipped into the ductwork as if custom-made. Joe replaced the vent without any damage whatsoever. Another perfect stash.

They had to get the name and address of the bank at the corner of the street. They had noticed it on their way to the house. Then Joe could make the call home to his parents and give them the information. They went out to grab a bite to eat and got the bank information at the same time.

Joe called his family in Montreal. His mother answered. There was a lot of "How are you doing?" and "What are you doing?" and "Where are you?" When Joe said he was in Germany, his mom told him that she and his dad were leaving for Paris, France, in a couple of days. They could all meet up in three days in Paris. Joe thought that could work out well. He told his mom that he would really like to see them but he had phoned because he needed some money. His mother understood and suggested that they meet at the Hôtel de Londres Eiffel in Paris in three days. When they saw him, they would give him some cash. Joe thanked her and told her he was looking forward to seeing them. Now, the boys had to get to Paris in the next couple of days. Paris was a nine-hour train ride from Munich. They only had that night and the next day in Munich. The following day would be traveling to Paris, and they would meet with Joe's parents the very next day. Holy shit. They went from no schedule to a stressful schedule.

In the meantime, when in Munich, one must go to a beer hall. They chose to go to the Hofbräuhaus, which prides itself to be the world's most famous beer hall. It dates back to 1589. The hall was beautiful, with waitresses dressed in traditional *dirndls*, and several oompah bands played while many patrons sang and swayed from side to side with the music. The hall served their own brewed beer in one-litre steins, with Bavarian food such as veal sausage with sweet mustard and pork roast.

Jack and Joe ate and drank like pigs. They befriended a group of German ladies that just happened to be sitting beside them. The girls were out for an evening of fun and celebration for the upcoming marriage of one of the girls. They must have toasted to the girl's happiness a dozen times, maybe more. And Jack and Joe participated in every toast. Everyone was pretty much wasted.

Before they knew it, the clock struck 2:00 a.m. Lots of people had left or were in the process of leaving. All the girls at their table had left to go home except for two, the two prettiest. Everything was meant to be. Their names were Gina and Lola. They were as pretty on the inside as they were on the outside. Jack mentioned to Joe that he hoped they didn't have a problem returning to their hostel at such a late hour. Gina overheard the boys talking. With a quick glance at Lola, Gina offered to put the boys up for the night. The girls lived together, so the boys could take one bedroom, and the girls the other. Of course, everyone knew that Jack was sleeping with Gina and Joe was going to be with Lola. Jack and Joe gratefully accepted their offer.

Everyone had some degree of a hangover the next morning. They all had a glass of orange juice and forced down some toast. It was Sunday, so the girls didn't have to go to work. They knew the boys had to leave the following day for Paris, so there was a mutual decision by everyone to just hang out together in the girls' apartment and play a board game. The boys were content with the plan but needed to check out of their hostel they never used and retrieve the hash. Jack suggested that he and Joe go back to their hostel and get their belongings. They could return to the girls' apartment later that evening and all go out for dinner, Jack's treat. They, of course, wanted to spend their final night in Munich with Gina and Lola and told the girls of their intentions. The girls seemed very pleased.

When the boys got back to their hostel, they didn't check out immediately. They wanted to pack up and get the hash first. They went up to their room. No one was around. Jack darted into the bathroom and closed the door. Joe unlocked their bedroom door and left it ajar. He then gave a quiet knock on the bathroom door as the signal for "All clear." Jack dashed out of the bathroom and into their room, carrying the sacks of hash in his arms.

Once they were safely behind their closed door, they decided to just put the hash in the luggage rather than put it in their boots. The chance of them getting stopped on the way to the girls' apartment was remote, and they were willing to take the chance.

Jack and Joe returned to the girls' apartment with their things. Gina and Lola had picked a restaurant for dinner and went ahead and made reservations for 7:00 p.m. The boys dropped their things, and off they went. Jack asked what they were having for dinner. Gina said they were going out for an authentic German dinner. She added that they were paying the bill because they were working and the boys needed to hang on to their money. Jack couldn't believe how lucky they were to have met such incredible women.

They got to the restaurant before seven, but the table was ready, so they were seated right away. Gina ordered four

Warsteiner beers to start. She and Lola liked that beer and wanted the boys to taste it. Jack commented how smooth it went down, not that just about any beer Jack ever tried went down well.

Jack knew that most German food was quite heavy and rich. There was meat served at every midday or evening meal, often at breakfast too. A typical meal usually included hearty meat portions drenched in rich, creamy sauces and always served with a full glass of beer.

The girls went over the menu, explaining what everything was. Gina was going to order a dish called Rinderroulade, or beef roll. Thin slices of beef rolled around bacon, pickles, and mustard are roasted with red wine, producing a dark, rich flabour. It came with a side of potato dumplings or roasted winter vegetables. Gina chose the potato dumplings as her side. Jack ordered the same thing but chose the winter vegetables for his side. Lola liked the schnitzel. The schnitzel was more of an Austrian dish, but extremely popular in Germany. At this restaurant, they offered the schnitzel with cheese and ham sandwiched within and served it with a green salad and potatoes. Joe really liked schnitzel and thought he would give that a try.

The waitress returned to the table to take their orders. They were ready and rattled off what they wanted. Before the waitress left, Lola ordered four more beers.

The meals came, and everyone enjoyed what they ordered. For dessert, they ordered four Spanish coffees and shared one slice of chocolate cake. Before they left, Jack ordered four shots of whiskey to go on a separate tab. When the shots came, Jack toasted the girls and thanked them for dinner. He ended the toast by saying how fortunate he and

Joe felt for having met two beautiful people. They returned to the apartment. Everyone couldn't wait to rip off their clothes and go to bed.

The next morning, the girls got up early to get ready for work. The boys had to get out by the time they left so the girls could lock up. They exchanged addresses and made a lot of promises. Jack really meant it when he said he wanted to see Gina again one day. After a lot of hugs and kisses, they left together and then went their separate ways. When the boys got out on the street, they asked the first passerby which way it was to the Munich Central Train Station. They were thankful that the instructions they got were right on. They arrived at the station and booked two seats to Paris, France.

Paris with the Parents

They boarded their train as soon as the conductor lowered the stairs, enabling the passengers to climb aboard. They had their choice of where to sit, being first on. They chose an empty cubicle and shut the sliding door. There was room for six people to sit down. There were bins above the seating for luggage, but six people and their luggage would be crammed, to say the least.

The train started to roll. It appeared that they wouldn't have to share their space with anybody else. Jack carried a traveling backgammon game in his bag. They played game after game. Jack and Joe were both quite proficient at backgammon. They always played for money—to make it interesting, but with very affordable stakes. It was about beating the other guy, mostly. And taking a few bucks from him too.

The first few hours on the train passed quickly. The next four or five hours passed slowly. They killed some time dozing off, but not really sleeping. Finally, they arrived in Paris, at the Gare de l'Est train station. As people got off the train from all the different cars, a line began to form to clear French customs.

They were next in line. When it was their turn, an officer asked them to place their luggage on a table against a wall. They didn't see any other passengers ahead of them put their luggage on the table for inspection. They placed their bags on the table, as asked. They had become somewhat accustomed to being treated differently, usually unfairly, because of the way they looked. An officer came over to them and asked some routine questions as he fumbled through their bags. Finding nothing illegal, he stamped their passports and sent them on their way. No problem!

Here they were, in "gay Paris." They decided to head to the Latin Quarter to find a place to stay. The Latin Quarter was a maze of narrow streets packed with tourists poking into all the funky shops. It was situated on the left bank of the river Seine, around the Sorbonne. As it was known for its student life, with several universities in the area, student demonstrations were always taking place in the Latin Quarter. The area had a lively atmosphere and many bistros packed with students and tourists.

They chose Hôtel Marignan. They had to meet Joe's parents the next day at the Hôtel de Londres Eiffel, which was only a kilometre away. Hôtel Marignan included a small kitchen, a library, and breakfast every morning. It was also the least-expensive accommodations of all the places the boys checked out.

Once they checked in, they went straight to their room. The first thing they did was take their boots off. Then went the hash. Carrying their luggage with boots full of hash was no easy feat. They were both bushed and fell asleep in their underwear. They slept all night like that. As a result, they got up with the sun. Before they could meet Joe's parents or do anything else, they had to find a stash. Vents proved to be a great hiding place, providing the cover could be removed undetected. The vent cover in their room had two decorative screws securing it over the vent. They removed the cover to expose the vent. The vent was round instead of rectangular, which made it a little more difficult to fit all the hash inside, but they did. As a result, the vent was virtually blocked. That didn't matter. What mattered was that their hash was safe.

It was about 7:00 a.m., still a little early to call Joe's parents and set up where and when to meet. It was a perfect time to take advantage of what the hotel considered breakfast. Down they went.

It was a help-yourself-type affair. There was coffee and tea, of course. The boys poured themselves a coffee and sat at one of the small tables in the room. There were a variety of breads laid out on a table, with a toaster sitting right beside the rolls. Jack chose rye toast, and Joe wanted the same, just very well done. Jack liked well done, but Joe preferred his toast almost burnt.

There was another table with two large platters, one with oranges and apples and bananas, and the other had a variety of muffins. Joe took an orange and an apple and a banana from the first platter, and two muffins from the second, a carrot muffin and a blueberry muffin. He brought it all back to the table and began cutting up the fruit into quarters, and each muffin was cut in half. That way, they both got a taste of everything. By the time Joe finished cutting everything up, Jack had the toast made and buttered. He grabbed some packets of jam and peanut butter and brought everything to the table. They were ready to eat.

It wasn't a five-star breakfast, but the boys totally enjoyed it. They had another coffee and watched other guests popping in and grabbing a piece of fruit and continuing on with their day. Jack and Joe were the only ones seated.

When they finished breakfast, they headed up to their room to make that call to Joe's parents. There was a phone in the room. Joe called, and his mom answered. There were the pleasantries before arranging to meet his parents at their hotel at noon. Joe's parents were going to treat the boys to lunch in their hotel. At 3:00 p.m., a bus was going to pick them up and take them on a tour, which would include dinner. Whether the boys had wanted to do the tour with Joe's parents or not, they had to go. Joe needed to get some money.

Jack and Joe arrived at Joe's parents' hotel just before noon. His parents were waiting for them in the lobby. There were hugs and chitchat before going into the dining room for lunch.

Everything in Paris had that "*je ne said quoi*," that European flair. Their hotel was no exception. They entered this large dining room and were seated at a table for six, even though they were only four people. Joe's father, Ralph, had asked the girl seating them if they could have a larger table to give them more room. Ralph was a fairly large man. Joe's mother, Honey, was a little bit of a thing. The lunch crowd hadn't come in yet, so they were able to accommodate his request.

Jack ordered *coq au vin*, a French dish of chicken braised with a Burgundy red wine, larders, mushrooms, and garlic on a bed of rice. Jack had eaten *coq au vin* at home before, made a little bit differently. Joe ordered the same thing as Jack. Joe's parents ordered the beef bourguignon, a stew braised in red wine sauce and flavored with carrots, onions, garlic, mushrooms, and bacon.

Both dishes were fabulous. They finished their cappuccinos and left the dining room. It was 2:00 p.m. They had an hour to kill until their bus picked them up for the tour. Honey suggested they hit the bar until the bus came. Honey had personality plus and loved to drink and cajole with the guys at the bar. Joe's dad, Ralph, was quiet and liked to sit back and watch Honey have a good time.

The bar was crowded with guests having a couple before lunch or having a couple after lunch, as they were. Honey, being a social butterfly, sat between two guys at the bar and began flirting with them within two minutes of them walking into the bar. Jack and Joe sat there taking it all in as they drank their beers, especially watching Honey interacting with all the men. Not one of those guys even knew Honey was with Ralph. Ralph didn't mind, so why should they? Everyone was having fun.

An announcement was made in the bar, to anyone who was waiting to be picked up, the bus had arrived. That was definitely their bus. Honey was a little tipsy, but everyone made it on the bus. There were about twenty other people from other hotels already seated when they got on.

The bus stopped outside a bar. Everyone got off the bus and followed the tour guide into the bar. Once they entered, they were ushered down to the stage where tables, closest to the stage, were reserved for the people on the tour. Everyone sat down and stared at the stage, wondering how they were going to be entertained.

Once everyone was seated, a guy wearing a burgundy suit came out on the stage. He introduced himself as Something the Great. Very vague. Nobody knew what he was going to do. He told the audience that in order to do his performance, he needed two volunteers. Two guys in the tour crowd put their hands up and were chosen. Something the Great, who was called Something for short, called the volunteers up on stage. They all stood onstage, making introductions and small talk. Something then snapped his fingers, and the stagehands brought out two comfortable-looking chairs. They were placed about eight feet apart, facing the audience. Something asked the volunteers to choose one of the chairs and sit down. The volunteers followed Something's instructions. Something snapped his fingers again, and the stagehands brought out what looked like two white sheets. Something took one of the sheets and covered the first volunteer as if he were in a barber's chair, about to get a haircut. He took the other sheet and covered the second volunteer in the same way. Something walked around the chair of the first volunteer as he talked to the audience about Paris and how he hoped everyone was enjoying his beautiful city. When all of a sudden, he ripped the sheet off the first guy. Everyone gasped. It caught everyone off guard. Something asked the volunteer to stand up. He did. Something asked the volunteer if he could see his wallet. The volunteer felt for his wallet, but it was gone. Something held up his wallet for the audience to see. Everyone clapped. Something asked the volunteer for the correct time. The volunteer looked at his wrist, and his watch was also gone. Something the Great held his watch up to the audience and everybody clapped again. Something gave the volunteer all his things back and thanked him for being a good sport. He was told he could sit back down, and as he did so, the whole room gave him a standing ovation.

Something started to walk around the second volunteer as he spoke to the audience again. Everyone in the audience was thinking the same thing: fool me once, shame on you; fool me twice, shame on me.

Everybody saw what he did to the first guy, so all eyes were on his every move. After a few minutes, Something tore away the sheet covering the second volunteer, and the same thing happened. The crowd went wild. Everyone stood as they clapped. Jack figured Something had probably pickpocketed them while he was talking to the volunteers before they even sat in the chairs. Any way you slice it, Jack though it frightening that there were people out there who could take your wallet without you knowing while dozens of people were watching.

When the second volunteer sat down, he, too, got a standing ovation. Something the Great thanked everyone, and he got a standing ovation with prolonged clapping. The next act, so to speak, was a guy name Bruno. He told the audience that, using mental telepathy, he could read any document held up by someone in the audience. All the volunteer had to do was to concentrate on what they were holding up in their hand, and Bruno would tell the audience what it said. Jack couldn't believe anyone could do that unless it was fixed somehow. No one volunteered. So Jack held up his hand to volunteer. Joe, Joe's parents, and everyone else in their tour group sat on the edge of their seats in anticipation of what Jack intended to hold up in his hand. Jack raised his passport up in the air, with the back of the passport facing the stage. Jack played the game and concentrated on his Canadian nationality and his name and where he lived.

Bruno appeared to go into a comatose condition. All of a sudden, he seemed to re-enter this world and said Jack was holding a Canadian passport. Bruno could see the colour blue, which was the Canadian passport colour. But a slew of other countries had a blue passport also. So it was quite impressive that he said a Canadian passport. Next, he told Jack that he lived in a city that began with the letter *M*. Now that got Jack's attention. The final thing he told Jack was that his last name started with a *D*. A moment later, he said to Jack, as if it had just come to him, that his first name was Jack. "Is that correct?" asked Bruno. Jack answered yes. Then he said, "Your home is in Montreal." Jack confirmed what Bruno had told the audience and sat down. Everyone clapped. Surely, there were people in the crowd who thought Jack must have been a plant. But Jack knew he wasn't a plant, obviously, and wondered how the hell this guy did it. Was it a very clever trick, or could he actually do what he claimed? Jack thought that if Bruno was really legit, he wouldn't be doing an act onstage; he would be a billionaire working at NASA or for the CIA.

Another man on the tour, Andy, raised his hand to volunteer. He was a fun-loving guy. Jack and Joe had been talking to him on the bus. Jack highly doubted that he was a plant.

Andy held up a sheet of paper. Anything could have been written on it. Bruno put his hand up to his head with two fingers of each hand pressing against his temples, trying to zone in on the piece of paper being held up. Bruno looked like he was deep in concentration, so the room was silent. You could hear a pin drop. And then Bruno said, "I'm sorry." Everyone looked around, wondering what they had missed. Bruno told the crowd that Andy was holding a letter stating that his divorce was now final. Everyone laughed. Andy yelled out, "It's a good thing!" The crowd laughed again. Bruno nailed it a second time. Jack thought there had to be a powerful camera somewhere. And the information would have to be communicated to Bruno. The only problem with that explanation was that when Jack held up his passport, it was closed. How could Bruno have gotten the information from the inside pages of the passport?

And that was it for that bar. Jack and Joe were quite impressed with the acts they had just seen. Everyone had to finish their drinks and get ready to move on. The bus was leaving for another club, Le Lido.

Le Lido was a cabaret and burlesque show. It was famous all over the world. They presented exotic shows with multiple dancers, singers, and other performers. Before the show started, they came around for drink and food orders. They all ordered the roast beef special, which they ate as they watched the show. The acrobat was sensational and funny at the same time. The dancers were spectacular. They took Jack's breath away. He had never seen anything so extravagant like that in real life, only on television. The show featured the Blue Bell Girls, the most glamorous and elegant dancers in Paris. Some say that the show put Vegas to shame. With all the lights and glitter, everyone on the tour left the Lido dazzled. It was a good way to end a great tour.

They left the Lido, and everyone got back on the bus. They were the last hotel picked up when they started the tour, so they were dropped off first at Joe's parents' hotel. Joe's mom remembered to slip Joe some cash. Jack thanked Ralph and Honey for a wonderful time and their extreme generosity. They were pleased that the boys had a good time. They all hugged and said their goodbyes and wished one another a safe rest of the trip. Honey tried to get the boys to tell her what their plans were, to no avail. The boys didn't have a plan to share. The only plan they had was to find a place in the sun by the ocean where they could kick back and walk in bare feet or sandals. Hopefully, they would find that place in Spain or just keep on trucking until they did.

As soon as they returned to their hotel, they went straight to their room and changed out of their Sunday best and into shorts and a T-shirt. The boys carried one pair of long pants other than their jeans, which they wore most of the time. They kept a clean shirt, just in case they needed to dress up. Everything was pretty wrinkled, but no one ever said anything, not that the boys cared.

They planned to leave for Barcelona, Spain, the next day. If they could get on a train in the afternoon, they could go to the Eiffel Tower in the morning before they left.

They got up early and loaded up the hash. They went down for breakfast and had the same breakfast as they had the day before. As they ate, they praised Joe's parents for the tour they had chosen. The food at the Lido had been out of this world, and the entertainment was top-notch. It was so thoughtful of them to include the boys on the tour.

After they were done with breakfast, they checked out of the hotel and headed straight to the train station to check out their options. They found out that they could take a night train to Barcelona. It was a little more expensive, but the boys thought it would be fun to sleep in a berth on a train. So they bought two tickets. The train didn't pull out until 11:00 p.m., so that gave them the whole day in Paris. Then, they could sleep on the train and get to Barcelona feeling well rested. Perfect.

While they were at the train station, they got a locker to hold their things, so they could gallivant around Paris freely without luggage. First, they had to go to the washroom and remove the hash from their boots and transfer it into their luggage. Then they stowed everything in the locker.

They left the train station and went to see the Eiffel Tower. The Eiffel Tower is a wrought iron lattice tower constructed to commemorate the centennial of the French Revolution of 1889 and show off France's industrial prowess at the same time. Being one of the most famous structures in the world, it is shown in just about every photo taken of the Paris skyline and almost every movie ever filmed in Paris. It's iconic. The French call the Eiffel Tower *La Dame de Fer*, the Iron Lady. You could walk up the stairs or take an elevator to the top and get a wonderful view of Paris. The boys took the elevator. Some people that were standing next to them in line chose to walk. They told Jack he should reconsider; taking the stairs would be memorable. They were wearing sneakers. The boys were wearing their boots. Jack answered that they were pressed for time but wanted to see the Eiffel Tower before they left France.

Apparently, there is so much to see if you do take the stairs that they offer tours, explaining the architecture and how the tower was built. The structure is all open to see as you climb to the top. They have pictures hanging on the walls of the different stages of construction. The boys missed out on all that. But there was no way they were going to climb the stairs in their boots. The elevator brought the boys up to the top in a minute, and they were able to get an incredible view of all of Paris. Now they were able to say that they went to the top of the Eiffel Tower when they were in Paris.

Paris is noted as the easiest walking city in the world, meaning that the city is so compact it is possible to walk to most attractions without having to take a bus or taxi. The boys were doing a lot of walking, but they weren't into waiting in lineups to see anything. There were lots and lots of things to do and see that didn't involve lineups. They did wait a while at the Eiffel Tower, but that was a must-see.

They avoided the packed cafés on the main drag that were often frequented by tourists. They found a café on a side street that had a nice little patio in a courtyard in the back. There were several people seated, some having a drink and others eating also. Jack and Joe were the only tourists there. All the other patrons seemed to be locals.

The waitress came over to their table. She spoke perfect English with a very sexy French accent. Jack ordered a full carafe of the house wine and a menu, using the best French he could muster. The waitress gave Jack a huge smile for his effort and left to fill the order.

The wine arrived, and Jack, in perfect French, told the waitress they weren't ready to order lunch yet. Again, she gave Jack a friendly smile and left. As they sat there drinking their wine, Joe mentioned that he had noticed that his ankle was bothering him and seemed sore to the touch, maybe the beginning of a blister. He had put a couple of Band-Aids over where he thought the blister was starting. Jack thought that was horrible news. They both hoped this blister thing wouldn't become an issue. They filled their glasses with wine and toasted to health, happiness, and success with you know what. The waitress came back to their table for the food order. Jack noticed the couple at the next table was eating the spaghetti Bolognese. It looked great. The couple saw the boys staring at their spaghetti and told them it was fantastic. Jack and Joe both ordered the spaghetti Bolognese with a side of garlic bread and a tossed salad.

A little more wine and their spaghetti appeared at the table. The couple next to them was right; the spaghetti was superb. They cleaned their plates and poured the rest of the wine equally into their glasses. When the waitress returned to take away their dishes, Jack told her how much they had enjoyed the meal and to bring them another full carafe of the house red wine. Everything was said in French, of course.

They still had several hours until they had to board their train. They had made friends with all kinds of Parisians. It was a very friendly place. Many, if not all, of the patrons knew one another. It was like one big happy family that had adopted Jack and Joe. Everyone spoke English and enjoyed listening to the boys' stories, the ones they could tell.

It was time to vamoose. The boys were feeling no pain. They said their goodbyes to the entire bar, including the bartender, and out they went. Joe's ankle wasn't bothering him anymore, probably because of the Band-Aids and the wine. Especially the wine.

Barcelona

They got to the Gare de Lyon train station and retrieved their luggage from the locker they had rented. They headed into the washroom to load up and get ready to travel.

They boarded their train. It was a seven-hour trip, arriving at Barcelona Sants train station about six in the morning. They made it to their berths, stowed their duffel bags, and waited to begin their journey into Spain. Jack assumed they didn't have to go through French customs again, because they were already on the train. They would only have to clear Spanish customs when they got off in Barcelona.

Jack saw a conductor and asked if they had to go through French customs, not wanting to have accidentally done anything wrong. The conductor told them that they would be going through French customs and Spanish customs when they got off in Barcelona.

The boys were all settled, so they decided to go out in the hallway and watch the people boarding the train, rather than lie in their berths. People were speaking French. People were speaking German. Some people spoke in languages the boys didn't recognize. No one spoke English. That is, until two very cute girls carrying backpacks boarded the train, speaking English. As they boarded, Jack was standing right there as if purposely there to greet them. Jack asked, "Where are you ladies from?" "Montreal," they both replied in unison, beaming with pride. Jack knew the pride they felt to be a Montrealer. Jack introduced himself and motioned with his arm in Joe's direction and said, "That's Joe." Jack told the ladies that they weren't going to believe it, but they were from Montreal also. If you meet someone on the road traveling and they are from your country, there is somewhat of an immediate bond. But meeting someone who is from your city is definitely bonding.

One girl introduced herself as Sally, and her friend was Bev. The girls were struggling a bit with their things, so Jack offered to give them a hand with their luggage and then help them find a seat. Sally and Bev were very grateful.

It was chaotic, but they found a couple of seats that were available with four other people in their compartment. The girls left their luggage on the seats and said they would be back. It didn't matter if anyone knew how to speak English; communication is nonverbal. Everyone saw them leave their luggage, so they knew they were coming back. Jack looked at the situation the girls were in and was thankful he had paid a little extra to get a berth.

There was absolutely no room for them to sit around and talk, so Jack suggested they go to the Club Car for a drink and talk and get acquainted. Everyone enthusiastically agreed.

When they got to the Club Car, the steward explained that they couldn't serve drinks until the train was in motion, but he allowed them to sit at a table and wait. They thanked him and sat down.

Everyone seemed to get along extremely well. Jack thought to himself that they were so lucky to have met a couple of fellow Montrealers. They were really pretty girls who were as happy to be with them to boot. He knew that Joe was thinking the same thing.

Sally, affectionately called Sal by her friends, seemed to gravitate to Jack, and Bev seemed to like Joe. The girls had flown to Italy and stayed with Bev's cousin for a week. They were backpacking it by train through France, also Spain, and ending up in Lisbon, Portugal. Sally had a cousin there she had been close to when she was a kid but hadn't seen her in ten years. They were going to bop around in Portugal for two weeks and then fly home to Montreal.

When it was Jack's turn to share their plans, he told the girls that they didn't have any plans carved in stone. They would probably spend two or three weeks in Spain and then return home also. Sal said they could have a reunion when they were all back home. Everyone smiled and nodded in agreement.

The Club Car was starting to get busy. They were all hungry, so they ordered four beers and the chicken sandwich with fries special. The sandwiches came quickly. It was definitely the most popular special, judging by the number of other passengers who had ordered the same meal.

Once they finished their meals, Jack noticed a huge lineup at the door of the Club Car, waiting to be seated. It was time to go. As they left the dining car, Jack led everybody to the boys' berths so the girls could see their accommodations. Sal and Bev were really impressed and wished they didn't have to go back to their seats with all those smelly people. Jack said, "No problem." He and Joe could sleep in one berth, and the girls could sleep in the other, hoping it wouldn't work out exactly like that. The girls graciously accepted.

Once it was decided that Sal and Bev didn't want to return to their seats, they all set out to retrieve the girls' luggage. They found their seats and were pleased that the luggage remained exactly how the girls left it. Jack grabbed one suitcase, and Joe grabbed the other. The girls waved to the people sitting there and left. They returned to the berths and stowed away the luggage.

The boys had to get the hash out of their boots without the girls seeing what they were doing. With his back to everyone, Jack took off his boots and placed the hash back into the empty boot, zipped up the boot, and stuffed his socks on top, hiding the hash. Joe saw what Jack had done, and when Jack was finished, Joe did the same thing. No one picked up on anything.

The rest of the night turned out the way the boys had hoped it would. It was a memorable time on the train to Barcelona. Two people in a berth, intended for one, as the train rocked from side to side constantly was an experience, to say the least. They even managed to get a couple of hours of shut-eye.

Before they got to Barcelona, they exchanged phone numbers and addresses and promised to have a reunion when they were all back home. The girls had to go through French customs too, so all four of them went together. Everything went smoothly. The girls went first and breezed right through and got their passports stamped. The boys followed right behind them, and their passports were stamped too. Then, the girls had to line up in a different line, because they weren't entering Spain. They were going straight through to Portugal. They wished one another well, with lots of kisses and hugs, and the girls were gone. Jack and Joe had to stand in the customs line for people entering Spain.

There were about a hundred feet between the French customs and the Spanish customs. The boys left the French border and walked the hundred feet to the Spanish customs. The Spanish border guy looked at their passports and refused to let them into Spain, citing they didn't have an exit stamp from France. For some reason, the French agent hadn't stamped their passport with an exit stamp. Jack and Joe had heard the sound of the stamps and had assumed their passports had been stamped properly. They never checked. The agent had obviously stamped his paperwork, but not their passports.

They were sent back to French customs to get their exit stamp. By that time, a new agent was working. They explained their situation to the new agent. He didn't know what to do, so he made a phone call to his superior. They were on the phone for a long time. Finally, the agent got off the phone and directed them to the second floor, where all the offices were located. He told the boys his superior wanted to see them. He was in office 202.

Off the boys traipsed. They went up to the second floor with a kilo of hash in their boots. There were agents all over the place, and several police supervisors had offices there. They walked along the hallway with every single person staring at them, wondering what they were doing there.

Finally, they came to room number 202. They knocked on the door. A voice from behind the door said, "Entrez!"

The boys walked in and told their story to the boss agent. He listened and then got on the phone with the Spanish customs. After he ended that call, he phoned downstairs to speak to his own agents. Everyone that was involved wasn't on duty any longer. It all got so confusing that the boss guy threw his hands up and told the boys to go back downstairs, and the French agent would stamp their passports with the required exit stamp.

Down they went to see the agent. He had obviously been instructed to stamp their passports, and he did. The boys headed back to Spanish customs. They were told not to create any more problems and stamped the boys' passports too.

The whole thing was ridiculous. Every time something weird like that happened, there was a chance the jig could be up. Something out of left field, absolutely impossible to foresee, could spell disaster.

They made it to Barcelona, Spain. That was sort of a mini goal. Part of that goal was to find a place to stay by the ocean. Hopefully, that would be in the cards.

In the meantime, they got a cheap room in a run-down hotel near Las Ramblas, a main street in the area where they were staying. Before they could go out to explore, they had to deal with the hash. All the vents were painted over with several coats of paint. Hiding the hash was going to be a challenge. They decided to put the hash at the bottom of their duffel bags, temporarily, until they could find a better place. And maybe they wouldn't be at that hotel much longer, anyway.

Las Ramblas

Las Ramblas is a two-way boulevard about one and a half kilometres long. There is a section towards the ocean that hails as Barcelona's red-light district, boasting the best strip clubs and brothels in the country. Jack and Joe had no idea of their surroundings when they checked into their hotel. So they were oblivious to what was in store for them.

During the day, tourists were in and out of the boutiques, souvenir shops, and eateries along Las Ramblas. And that was exactly what the boys did. They bought some souvenirs and had a great lunch in one of the many restaurants. At night, the area transformed into a wild party with hundreds of sailors, on leave from their ships docked in the harbour. They had lots of money in their pockets and a burning desire to have a good time. Legal prostitution flourished, and the bars and brothels were packed. Three American dollars could get you anything you wanted.

Not knowing what was going on, Jack and Joe walked into a bar to have a drink and something to eat. No one else was there except for the bartender, no other customers. When they walked in, they smiled at two very friendly waitresses who seated them in a booth. As soon as the boys sat down, one waitress sat down beside Joe, and the other beside Jack. They asked what the boys wanted to drink. Jack said two Heineken beers. The waitress yelled out to the bartender, "Two Heinekens and two cognacs." Before the words were out of her mouth, her hands were on Jack's family jewels, and Joe had the same thing happening across the table.

Their old philosophy of not paying for sex was ridiculous. Jack always thought that people who paid for hookers were desperate, not being able to get a woman without paying. And perhaps that was true of some of the people who picked up street hookers, but the women in these brothels were professionals. For healthy men in their twenties, Jack and Joe would be insane to adhere to that policy. Friends back home only dreamed of being in their situation.

The boys had gone into the restaurant to get a drink and something to eat. They were caught totally off guard and perhaps a bit intimidated by the waitresses. So they drank up and excused themselves, much to the ladies' astonishment, and paid for their beers and the girls cognacs too. The boys only had one rule: avoid trouble, no matter what.

They tried to get a drink in another bar, and the exact same thing happened. This time, they left without ordering. When they got back onto the sidewalk, three drunken sailors were passing the bar on their way to their next conquest. Everyone bumped into one another. There was no threat of a fight, only friendly, good vibes. So much so that after some sincere drunken apologies, one thing led to another and they all began to hang out together. They ventured out into the heart of the red-light district. The sailors were bragging about what they had seen and what they had done while on leave from their military duties. They mostly bragged about how many different women they had been with.

One of the sailors, Al, was sort of the leader. He asked Jack if they had run into anything special. Jack said they had just gotten into town and they hadn't been with any women. The sailors laughed in disbelief. The sailors said they wanted to take Jack and Joe to a great place they had found, and they would pay the bill. How could anyone refuse a deal like that?

They arrived at this club and went in. They were greeted by this very friendly lady at the door who welcomed them in. Al told the lady that they were a party of five. She led them into a room with a bar and several plush sofas and chairs upholstered in bordeaux-coloured velvet. All the available girls entered the room and said hello. Al was the first to make his choice, and off he went. The other two sailors followed Al's lead pretty quickly. Jack and Joe were a very close second behind them, not wanting to be embarrassed. It was incredible, from the choosing of the lady to the very end of the evening.

Jack and Joe met the sailors outside the club when everyone was done. The boys thanked the sailors for a really good time. It was unbelievable that total strangers would treat Jack and Joe to such an awesome night. The sailors were pleased. It was three o'clock in the morning. The sailors kept on going, as sailors do. Jack and Joe bowed out and headed back to the hotel. When they got back to their room, they stripped off their clothes, jumped into a shower, and hit the hay. They were beat, especially after all the excitement of the evening.

The next morning, they decided to stay at the same hotel for another night. They couldn't leave Barcelona without seeing a bullfight. Jack and Joe chose to go to La Monumental to see one.

The night before, they had left the hash in their duffel bags when they went out on the town. They had stacked their duffel bags end to end, with the zippers lining up and facing the wall. Anybody who moved either bag would never know how to put them back exactly the same way. The boys figured, if anyone looked in their duffel bags, they would be looking for valuables and would probably overlook the hash. Their alternative was to carry the hash on them, which was out of the question.

On the way out of the hotel, Jack informed the lady at the front desk that they would be staying for another night. After breakfast, they headed to La Monumental to buy tickets to see a bullfight.

The Bullfight

The boys were delighted that the bullfight was starting in less than one hour. They paid for their tickets and went searching for their seats so they would be ready for the start of the fight.

Spanish bullfighting was practiced in Spain, Mexico, Columbia, Ecuador, Venezuela, Peru, and part of southern France and Portugal. The sport is to publicly subdue, immobilize, or kill a bull, usually the latter. The most common bull used was a Spanish fighting bull, called a Toro Bravo, a type of cattle native to the Iberian Peninsula. The bullfight is both a sport and an art performance. The red colour of the cape is only a matter of tradition, because the bull is colour-blind. They attack moving objects, and the red-coloured cape was chosen to mask the blood. It's really gory!

In a traditional corrida, or bullfight, three matadors each fight two fighting bulls to death. Each bull has to be at least four years old and must weigh a minimum of one thousand pounds but can weigh up to thirteen hundred pounds.

The start of the fight was announced by a trumpet sound. It was exciting. The participants entered the area in a parade accompanied by band music playing *pasodobles* composed to honour famous *toreros*, or bullfighters. Matadors were distinguished by a custom-made "suit of lights," embroidered with silver or gold thread.

The bull entered the ring to be tested for aggressiveness by the matador. He observed how the bull reacted to the waving of the cloak. He also noted vision problems, unusual head movements, or if the bull favoured a particular part of the ring, which is called his querencia, or territory. A bull trying to reach his querencia was often more dangerous than when he was charging the matador.

Then the matador began his performance with several lances, or passes, showing control over the bull. He progressed to the fancier Veronica lance, which was when he let his cloak trail over the beast's head as it ran past. And the crowd went wild when they saw him do that.

Then, two picadors entered the arena armed with their long lances and mounted on large, heavily padded, and blindfolded horses. The bull was attracted to them when they entered the arena and immediately charged at one of the horses. In most bullfights, when the bull lowers its head to charge at the horse, the picador would stab his lance into the mound of muscle on the bull's neck, blood gushing like a fountain from his wound. That's the way it is supposed to happen. In that fight, the bull ran into the horse, knocking the horse over and sending the picador flying through the air. Jack and Joe got up and cheered. Of the fifteen thousand people in the arena, Jack and Joe were the only ones that got up and cheered for the bull. The picador's legs were in plaster casts for his protection, so once thrown off his horse, he couldn't walk. He had to just lie there until the clowns ran out and distracted the bull so the picador could be carried off to safety.

Then came the banderillas. The matador planted two barbed, dart-like sticks, known as banderillas, into the bull's shoulders. That further aggravated the bull and weakened the ridges of the bull's neck and shoulder muscles. By that time, the bull had lost a significant amount of blood and was beginning to succumb to its exhaustion. The matador re-entered the arena with a small red cape in one hand and a sword in the other. A clumsy kill that failed to give a swift and clean death would often raise large protests from the crowd and could even ruin the entire performance for some. There were rules in place about how to kill the bull, and safeguards against the bull suffering. If the Presidente considered the bull fierce enough, he could spare its life and retire the bull, allowing him to be put to pasture to stud. However, that was rare. A bull was never used twice, as it could actually learn from the experience, if it survived. If the Presidente appreciated the fight, he would award the matador with one ear, both ears, or both ears, and the tail, depending on how the Presidente judged the matador's performance. After the bull had been killed, it was picked up and butchered. The meat was then distributed to the needy. Jack and Joe were glad they could say they had seen a bullfight firsthand but thought it was barbaric. It was something they did but had no desire to ever do again.

They found a restaurant that looked inviting, where they could get some lunch. There were a few tourists, but it wasn't overrun like some of the other places they had passed. They ordered Botifarra, a white pork sausage with mixed vegetables on a bed of rice. To drink, they stuck with Heinekens.

The boys weren't into seeing sights and waiting in long lineups. They were more enthusiastic about finding a place to stay by the ocean, and they both had a feeling that they were in the right vicinity.

When paying the bill, Jack asked their server if she could suggest a place for them to stay by the ocean. She was proud to be a Barcelonina and suggested they take a ferry to the island of Ibiza. It was a nine-hour ferry ride through the Balearic Sea to San Antonio, Ibiza. She said they would enjoy the beauty and the nightlife. She came up with an alternative solution too. She suggested they could jump on a train and travel approximately thirty-five kilometres outside of Barcelona to a little town called Sitges. It was right on the Mediterranean Sea, and the nightlife was great there too. The boys thanked their server for all her help, tipped her handsomely, and left.

There wasn't much discussion about their next move: Jack and Joe wanted to go to Ibiza. They headed down to the ferry dock and inquired about the cost and schedule to San Antonio, Ibiza. There were two trips every day, one leaving at 10:00 a.m. and the other at 11:00 a.m. Advanced reservations weren't required. You just showed up and bought a ticket. Perfect. They decided, on the spot, that they would be on the ten o'clock ferry the next morning.

Back to the hotel they went, with a bit of a jump in their step, being very excited to be going to an island off the Spanish coast. This just might be the island in the sun that they had been looking for.

It was time for a little siesta before dinner. It turned out to be more of a lay-down-and-rest while discussing plans for after dinner. It was their last night in Barcelona, so they decided to get something to eat on Las Ramblas and deal with whatever came their way after that.

They cleaned up and left for their last night in Barcelona. They walked down the street, taking everything in—the bright lights, the laughter, the beautiful women. Everyone was having fun.

They spotted a bar that sold hamburgers, or a hamburger place that sold beer. They sat down and ordered what was advertised as a world-class hamburger with local fries, *papas bravas*, fried potatoes in a spicy sauce, which were very popular with Barcelonians. For drinks, they stayed true to Heineken and ordered a couple. With a last-moment change of thought, Jack called back the waitress and cancelled the two Heinekens, asking her to bring them two of whatever was most popular in Barcelona. She said Estrella Damm. They said fine. A lot of people drank Estrella because it was brewed locally. Sitting on the patio of the hamburger joint, Jack and Joe watched the people walking by as they drank their beers and ate their burgers and fries. When they finished their burgers, they ordered another round of beer. They were both very content to just sit there and drink beer as they watched the action on the street. It was exciting.

A bunch of sailors sat down to eat a burger. They ordered some beer, and everybody became sort of friendly. The sailors started telling the boys hilarious typical-sailors-getting-into-trouble-on-leave stories.

It was almost midnight, and the boys were drunk enough to make very loose decisions. Joe had seen a girl in one of the bars the night before that he had fallen in love with. He suggested that they go back to the bar to see if she was working, so he could spend some time with her before they went back to the room. Off they went to find Joe's love. They spent a couple of hours there at the bar after finding her, Joe with his love and Jack with her amazing friend. They brought Jack and Joe to a room in the back of the building, where they had some privacy and were able to relax and have fun.

They got back to the hotel around 2:00 a.m. They needed to be at the ferry by ten o'clock the next morning. As soon as they hit the pillow after showering, they were gonzo.

Ibiza and Formentera

Morning came quickly. Jack was in the shower by 8:00 a.m. Joe did his thing next. They loaded up their boots with the hash and headed out to the ferry dock. On the way, they passed a fruit stand and purchased six bananas, six oranges, and six apples. That was their breakfast and snacks for the trip to Ibiza.

The captain came out and lowered the chain on the gangplank, opening the large boat to the waiting passengers. Jack and Joe were first in line. There were only a few people getting on the ferry, but there was still twenty minutes before departure. The boys found two seats towards the bow and shoved their duffel bags underneath the chairs. As they waited to set out, they watched the last of the passengers scramble to get aboard. The very last of these stragglers were three young guys. They were traveling with backpacks, but Jack thought they didn't seem like the type to be backpacking through Europe, for whatever reason. It was just a feeling deep in his gut.

The ferry left the dock at precisely 10:00 a.m. If you were one minute late, it was the 11:00 a.m. ferry for you. There was no waiting. Most of the passengers sat on the top deck, watching the Barcelona skyline fade and then disappear. It was a gorgeous day, so a lot of people on the top deck were taking the sun, while others enjoyed scanning the water for whales or dolphins. The lower deck was attractive to all those who wanted to stay out of the sun and was used by everyone when it rained or if the temperature dipped.

It wasn't long before Jack and Joe crossed paths with the three guys who had made it onto the ferry by less than a minute. Introductions were made. Good vibes were felt by all, but Jack still never let his guard down. Their names were Sonny, Willy, and Junior. They said they were traveling around for the spring and summer. Jack said they were doing the same thing. But Jack knew that Sonny was lying.

They had a long time to kill until they reached San Antonio. The boys hung out with their newfound friends for the entire trip, mainly because there was no one else whom they could relate to on the ferry. During all those hours, they became friends of a sort. Sonny confided in Jack that the truth was they were not traveling around for the summer, not as tourists. They were professional pickpockets. They went from fair to convention to anywhere large crowds gathered. They could steal a wallet or a handbag without the person knowing they were robbed until they were out of sight. The only good thing that Jack learned from Sonny was the fact that they always took the easy score. They looked for the tourist who was careless, the one not paying attention. If you remembered to keep your wits about you at all times, you were a bit safer from those bad dudes. It sounded basic, but there were so many tourists just having a good time on vacation who let their guard down and became vulnerable. Jack was glad he didn't make a living that way. Jack wasn't being judgemental; he just personally couldn't ruin people's vacation for his gain. Jack felt it was bad karma.

Jack had known at first sight that Sonny and his friends weren't what they said they were. Jack reminded himself to always take note of his gut feelings. That feeling was never wrong. But the boys couldn't care less if that was what they did for a living. It was their karma. Jack knew if Sonny or his buddies ever saw the hash, they would relieve them of it in a flash, friends or no friends.

Finally, the ferry arrived in San Antonio. Sonny and his crew were taking a few days off from work. They had done really well at an antique car show in Barcelona earlier that week and had decided to travel around for a few days. They had never been to Ibiza before and had no plans whatsoever. Jack and Joe were in a similar situation with no firm plans, so they all continued to hang out together. They began looking for a place to stay to accommodate all five of them. Jack smelled problems. They were in no position to have anybody hanging around, especially these guys.

They found a hostel for all five. It wasn't separate rooms with privacy; it was a large room with six bunk beds. There wasn't an opportunity to find a place to hide the hash, so they just kept their boots on.

Everyone was hungry. They decided to go out and find a nice place to eat. Sonny, Willy, and Junior started changing out of their sneakers into their sandals. Jack and Joe would have loved to do the same but couldn't. Junior asked why they didn't want to change out of their boots. Jack lied and told him they had ruined their sandals a couple of weeks earlier and were going to purchase new ones at their first opportunity. Everyone seemed to accept that, but Jack caught a weird look on Sonny's face. Sonny was no dummy; he made a living reading people.

Off they went. They found a place by the water that looked perfect. They all sat down. The food smelled great, and they were all very hungry.

The waitress came over to take the order. They ordered five of whatever smelled so good. "Sofrit pages" was the daily special, a traditional Belearic stew with a hearty mixture of various meats like chicken, lamb and Ibizan sausage, sobressada, and botiferra. Estrella Damm was popular on Ibiza too, but Jack ordered five Cruz Campos, just to try another popular brand.

As they waited for their grub, Sonny mentioned checking out the island of Formentera, a thirty-five-minute ferry ride from Ibiza. Jack jumped all over that idea. Their present situation at the hostel had to change, and going to Formentera would do just that.

Everyone drank up, and they returned to the hostel. As soon as they got back, Jack grabbed his toothbrush and announced that he was going to hit the can before crashing. When in the bathroom, he did his thing and then took all his clothes off except for his T-shirt and underwear. He rolled the hash up in his clothes and returned to the room. As soon as Jack got back, Joe knew what he had done with the hash and said, "Next," and headed to the bathroom before anyone had a chance to object. Jack put his clothes, with the hash rolled inside them, into his duffel bag, and no one was the wiser. Soon, Joe returned and did the same thing. The next morning, they would have to do an encore, only in reverse.

Jack was the first to get up. Sonny and his buddies were dead to the world. Joe was just waking up and saw Jack go to the washroom. He knew he had better be ready to go next. When Jack returned, Joe hustled into the bathroom and got organized too. By the time he got back to the room, Sonny, Willy, and Junior were up and dressing for the ferry ride to Formentera. Again, there were weird looks from all three when they noticed Jack and Joe wearing boots instead of sandals, even though they had explained why. There were no comments made this time, but Jack knew it was only a matter of time until Sonny, especially, figured something was up. Jack knew Sonny was suspicious and might have already been planning to rip them off. Sonny might have thought that they were hiding money or something of value in their boots.

They arrived at the ferry dock with thirty minutes to spare. There was a store that sold a variety of different things a stone's throw from where they were boarding. Everyone bought a couple of pieces of fruit, one to eat for breakfast, and the other for the trip. The store had a baked goods section. They got five wonderful-looking croissants and several slices of a white cheese they had sampled and really liked but never got the name of.

As they ate their breakfast, Sonny noticed four tourists, two couples, whom he determined were easy prey. While they conspired to relieve these poor, unsuspecting people of their wallets, Jack and Joe finally had a moment to themselves to have a private conversation. They had both decided independently that it was time to move on by themselves or risk imminent discovery. That in itself was a huge problem, but now, Sonny and his buddies were committing crimes in their company. If those guys got caught, it would be lights-out for Jack and Joe too.

They decided to follow through with the ferry to Formentera, mostly to avoid any problems with Sonny. When they got to Formentera, they could find an excuse to go back to Barcelona. Joe wanted to go to Morocco, just across the Strait of Gibraltar. Jack wanted to check out Sitges, that little town thirty-five kilometres outside of Barcelona. It was closer. If that didn't work out, they would travel to Morocco. It was agreed.

The trip to Formentera went quickly. It was only thirty-five minutes on the ferry to their landing at La Savina. All ships, boats, and ferries docked there. When they got there, it was almost desolate. There was only one hostel, and nothing else. Other tourists took a taxi into town to stay at small privately run hotels. The boys weren't into doing that with these guys. They just wanted to return to Barcelona. So they got their accommodations together again, right near the ferry, for an easy getaway. They were able to get two rooms, so that alleviated one big problem. At least now they had some privacy. Jack and Joe were going to tell Sonny that evening at supper that they were returning to Barcelona in the morning. Second problem almost solved.

Formentera was a naturalist's paradise. There were no tall buildings. They had beautiful, clothing-optional beaches. There was no camping. Biking was very popular with most people who visited there, riding through gorgeous scenery right alongside the ocean. It was ideal for those who enjoyed long walks along the beach or hiking through the woods.

The hostel had a snack bar. They sold sausages and all kinds of junk food. So they all had a sausage and a bag of chips. Jack suggested that they go into town for supper, where they could sit down and have a proper meal and see a little of the island. Everyone agreed that sounded like a good idea. Jack was setting the scene to tell Sonny that they were leaving in the morning for Barcelona.

They got the people at the hostel to call a taxi to take them to the town of Portu Saler, seven kilometres away. There was a bus service from the ferry dock to the village if any tourist wanted to skimp on the taxi fare to Portu Saler or wanted the experience of riding the public transportation.

The taxi arrived, and they all piled in for the trip to town. The town was small, and there weren't many bars or restaurants, but they found a nice place where they could sit on a patio and have dinner and a few brews. There was even a live band playing music. The guy singing would walk off a stage of sorts—it was more like a plywood platform—and pick out a couple sitting at a table. He would sing them a love song, hoping for a tip. And he always got one, the couple feeling it was so romantic and eating up the whole thing.

Everyone finished eating and was working on their third Estrella. Jack thought the time was right to let Sonny know their plans had changed. He came right out and made the announcement that he and Joe had decided to abandon their plans of seeing all the Balearic Islands. They were returning to Barcelona in the morning. Before Sonny could ask why, Jack lied again and said they had left a bag at their hotel in Barcelona that had some important papers they didn't want to lose. It pissed Jack off that they had to change their plans, but getting as far from Sonny and his friends was paramount.

Sonny reacted positively and even seemed relieved. Jack and Joe couldn't wait to part company with them, but it turned out they thought Jack and Joe were weird and were happy to move on separately. Go figure.

They all toasted one another with the remains of their Estrellas, wishing one another success in their future travels. There was a taxi sitting right outside the café, looking for a fare. They all jumped in and returned to their hostel. The boys went to their room, and Sonny and his friends went to theirs.

In the morning, everyone said their goodbyes, while all thinking, *Good riddance*. Jack and Joe went straight to the ferry dock and were prepared to wait for the next ferry. They hadn't even checked the schedule. It didn't matter. They were "just them" again.

Sitges

It was a short ferry ride back to Ibiza. Ibiza was beautiful. Spending any time there was out of the question, though. They could easily bump into Sonny and crew again. So they decided to keep trucking on to Barcelona. They were fortunate they were able to connect to the ferry going to Barcelona with only a twenty-minute wait. It was a nine-hour trip to Barcelona, so they found a place where they could get some sandwiches and snacks. A couple of pieces of fruit hadn't been enough on the way to Ibiza. They remembered being really hungry.

By the time they got back to the dock, they were able to board. They found a place to sit, secured their bags under the seats, and went the six feet to the railing and watched all the people scrambling around, boarding the ferry. They continued to stand there and watch the ferry pulling away from the dock and Ibiza becoming smaller and smaller, until the island disappeared.

It was a beautiful day, as always. Jack couldn't remember the last time it had rained. Joe took out his traveling backgammon board, and they played for hours. At one point, they paused the game to watch a number of dolphins putting on a show for them. They also saw a swordfish leap out of the water and dive back in, right in front of them.

They made it back to Barcelona around five o'clock in the afternoon. They had two options: they could go straight to Sitges or spend another night on Las Ramblas, which wouldn't be the end of the world.

They decided to jump on a train and sleep in Sitges. They would have to go back to Barcelona on the way home, anyway, so they justified their decision to pass on another night there and go straight to Sitges. They hurried to the train station to see if it was possible to get to Sitges that evening. The last train was leaving in ten minutes. There wouldn't be another train until six the next morning. They bought two tickets and boarded. It was only thirty-five kilometres to Sitges, and the train ride took about half an hour.

When they arrived in Sitges, it was just after six. They inquired about getting to the nearest hotel. They were informed that there were absolutely no vacancies in any hotel or hostel in the town due to some convention that weekend. They sat there on their duffel bags, pondering their next move. Anybody else in their situation could have slept on a park bench. But the boys couldn't risk being arrested for loitering. They couldn't go to the police and explain their situation, because that would likely involve taking their boots off at some point. The sun was going to set in a couple of hours, and then what? They were in trouble. Joe mumbled under his breath that they should have stayed in Barcelona.

They sat there totally deflated as the sun began to set. They would have to find a quiet area where they could get through the night, constantly keeping an eye out for patrolling police. The next day, they would make a major effort to find a place to stay. It was depressing.

The boys had to move. All the other passengers were long gone, and the boys stuck out like a sore thumb. All of a sudden, Jack felt a tap on his shoulder. He turned, expecting to see a police officer instructing them to move on. Instead, there stood a wizened-up old lady dressed entirely in black. She spoke no English but put her hands together as if to pray and rested them beside her head and tilted her head onto her shoulder. Jack immediately understood and replied, "Si, señora, si." She motioned with her hand to follow. They did. There they were, following this strange lady to who knew where. But they definitely didn't have anything better going on. They were desperate.

The boys followed behind her with their bags slung over their shoulders. They were looking at everything around them as they walked, taking it all in. Every few minutes, the little old lady turned around to see if the boys were still following and gave them a hand motion for "Come on." She brought them to this street that had the ocean on the left and these big old houses on the right, on higher ground, with a view of the ocean. Eventually, she stopped in front of one of these large houses and pulled out a key and opened the front door. She went inside and motioned for the boys to follow.

There was a vestibule as you walked in, with a door on the left and a door on the right, and a steep, steep staircase to the second and third floors straight ahead. The old lady pointed to herself and motioned with her hand up to her mouth as if to eat, so Jack knew that was her kitchen and living space.

They climbed—and *climbed* is the appropriate word—up the staircase to the second floor. There was only one archway leading into a large room. There was a big long dining room table and chairs. All the other furniture was draped in plastic. It was obvious that the room was seldom, if ever, used. On the wall behind the table, there were framed pictures of this little old lady posing proudly with people of all different nationalities and colours. If you saw pictures on the wall of family, that was fine. But this little old lady sitting with such a mixture of friends and family seemed very strange to Jack.

Without ever really going into the room on the second floor, just a look in, she ushered the boys to the staircase and up to the third floor. At the top of the staircase was a large skylight, probably lighting the staircase with much-needed light during the day.

To the left was a washroom. The door was ajar, so they were able to see that it looked clean and well-maintained. To the right was a double door leading into a room with two double beds. It had a large table against one wall, and a step up into an open kitchen with a working fridge and a hot plate. She opened a cupboard and pointed out plates and bowls, and in a drawer was cutlery.

Jack and Joe felt like they had won the lottery. Jack made a motion with his fingers as if handling money and asked, "Cuanto?" All in sign language, she pointed to Jack and held up five fingers. Then, she pointed to Joe and held up five fingers. Jack wanted to confirm that it was per day, so he held up one finger but assumed it couldn't be for one week. She put up seven fingers. She was charging them five bucks each per week for being in this house opposite the ocean. Of course, they agreed and thanked her over and over. They shook hands and introduced themselves. The little old lady's name was Maria.

This was what they had been looking for. Jack lived by the phrases "Everything is meant to be" and "Everything happens for a reason, even if it isn't clear at the time." If they hadn't gone through that Sonny thing and changed their plans, they wouldn't be here now. And above all, always, "Keep the faith." There was always a pony under the pile of shit. A positive attitude is magic. And this was proof that all that was true.

That night, once they got settled, they walked down to the ocean. They walked along the beach and found a perfect place to take the sun, where apparently many locals went to swim. That was definitely the plan for *mañana.*

When they got back to the house that night, the front door was unlocked. Jack was thankful, because Maria hadn't given them a key. They were extremely quiet, not wanting to disturb her. They mounted the hill of stairs to their room on the third floor. They had left the hash in their duffel bags before they had gone out, thinking it would be pretty safe. Now they had to find a stash.

The room had a vent. The cover came off easily, and the hash fit in the vent like a glove. That done, they hit the can and went to bed. As they lay there in their beds, they talked about what a topsy-turvy day it had been. One minute, they were frightful of their situation, and the next, they were thankful.

They got up the next morning and went out for breakfast. They wore shorts, a T-shirt, and sandals for the first time in months. On the way out, they bumped into Maria. Jack hadn't given Maria any money yet, so he took care of that. She nodded her thanks and grabbed Jack's arm lightly and led the boys into the room on the second floor. She orchestrated the scene without saying a word. She pointed to a picture on the wall of these two guys with her in the middle. She gently pushed Jack into one chair, and another gentle push for Joe in another, and she sat in the middle. Once seated, Maria yelled down towards the first floor something that sounded like "Tally."

Almost immediately, the boys heard footsteps running up the stairs. An elderly woman the boys had never met walked into the room carrying a camera. Maria began instructing her friend Tally in Spanish. Tally made a motion to smile, and they all did. Tally took their picture.

Well, that explained the pictures Jack had seen the night before of Maria and all those people she was posing with. They were pictures of people who had stayed with her over the years. Soon, their picture would become part of Maria's collection too. Maria was so proud. She hugged the boys before they left.

This was the boys' first glimpse of their neighbourhood in the daylight. As soon as they walked out the door, they smelled the ocean air. Right on the corner of their street was a café with a patio overlooking the ocean. That looked like the perfect place for breakfast. They sat down at a table on the patio, mesmerized by the sound of the waves crashing into the shore. They felt like they had arrived.

The waitress came over to the table to take their order. Her name was Sarah. They ordered two coffees and two carrot muffins. They sat there drinking their coffees for about an hour, cajoling with Sarah at every opportunity. Jack knew that wouldn't be the last time he saw Sarah.

They left the café and walked down the main boulevard to the beach. They hadn't brought their bathing suits, so they took off their sandals and walked barefoot through the shallow water along the ocean's edge. As they walked, they decided to return the next day prepared to take the sun and go for a dip in the ocean. They walked for hours, first along the entire beach, and then through town, up one street and down another, noting which stores and restaurants they would go back to another time.

They realized there was a lot to see in Sitges. The town dated back hundreds of years and was known for its historical sites, as well as its beaches and nightlife. They had just missed the Sitges Carnival, famous for its colourful floats, fancy costumes, and wild after-hours parties. Jack and Joe were happy they had missed it, the population of the town quadrupling for that week. It made it impossible to go to a restaurant or anywhere else in town without waiting in a lineup.

It was almost time for dinner. They grabbed a bite to eat and then went back to their room and relaxed for a while before checking out the nightlife. Apparently, nothing started until much later in the evening, so they probably had a little downtime after all that running around.

They got back to their room at Maria's house. The boys thought it was a good idea to smoke a joint before they went to the bar and maybe roll a couple for later. The last time they had smoked a joint was on the kibbutz in Israel. They hadn't dared to smoke a joint since. Now that they were somewhat settled, there was no problem walking down to the ocean and smoking a joint.

Joe removed the open sack of hash from the vent and purposely didn't replace the vent cover. They decided to roll several joints, since they already had the hash out. They started a production line. Joe heated the hash with his lighter so he could crumble it up. The hash became supple with the heat, and Joe began to make a small pile of crumbled hash. They broke up several cigarettes and made a pile of tobacco. Jack took the crumbled hash, mixed it with some tobacco, and rolled the joint. Then Joe added the filters. They continued until they had an entire cigarette pack full of hash joints. They put the remainder of the hash back into the vent and replaced the cover. They were ready to experience Sitges at night.

As planned, they left their abode and walked down to the ocean to smoke a joint. It was really dark, and no one was around. The boys passed the joint back and forth as they rambled on about their good fortune. They had just taken the last toke when a couple arrived in the pitch-dark. The couple was totally shocked to see someone lurking in the dark. Jack saw a look of fear on their faces. Jack and Joe hadn't expected anyone coming down to where they were either. The boys said hello and scooted off to the bar. The couple was relieved to see them take off so quickly and carried on with their romantic encounter.

Sitges wasn't a large town. You could pretty much walk everywhere. The day before, the boys had enjoyed a beer at a bar just up the street from their room. They got good vibes from the staff and patrons. It was a good bar to start with because they had gotten to know some of the locals who hung out there. They could easily try another bar if that one didn't tickle their fancy.

The bar was dynamite, as expected. There were a lot of ladies who were on vacation and out to have a good time. The boys met several interesting people. In general, people on vacation were always fun, because they were happy.

The boys stayed at the bar until last call. They said adios to everyone they'd met, threatening to do it all again the next evening. They went down along the ocean on the way back to their room and pulled out a joint. They were a bit tipsy from the booze, and after passing the joint back and forth a few times, they began to make each other laugh. They laughed until their stomachs hurt. Every time that happened—and it did quite often—they both commented on how good it felt.

They got into a routine of going to the beach in the morning and early afternoon to work on their tans. They became very tanned with all their effort. They would choose different restaurants with a patio and have lunch. In most instances, they had lunch and got drunk. Everything was conducive to having fun. Tourists, after a few drinks, were usually very friendly people.

Sometimes, they made it back to their room for a nap. Other times, they went straight from the patio to the bar. Once they got to the bar, they nursed one or two beers but snuck out every hour or so to smoke a joint. Nine times out of ten, they met a nice lady and went to her room or even started to bring their friends home to Maria's house.

They had been in Sitges for over a month. Jack was running out of money. They were spending a lot of money on partying even though things cost a fraction of what they cost back home. Jack had to ask Pops to wire him five bills. He went to the bank and got their address and notified the bank of the address where he was staying. Then, he phoned Pops and arranged to send the funds to that bank. Jack told his dad they were doing well and would see him soon. And of course, that he loved him. Jack thought to himself, he better not get busted. Pops would be mortified. Jack would be too.

In a couple of days, Jack had his cash. He intended to pay his dad back from their profit, after selling the hash.

Sister Ruthie

One morning, Jack was lying in bed, still half-asleep, when he heard his sister Ruthie's voice. He shook the cobwebs from his mind. He wasn't dreaming. Nobody in the world knew that he was living in this old lady's house, so that couldn't be his sister Ruthie's voice. It was impossible. The more Jack lay there, listening to this girl trying to communicate with Maria, the more Jack knew that was Ruthie.

That was enough. Jack jumped out of his warm bed and threw some clothes on before going out into the vestibule to investigate. As Jack opened the bedroom door, he was face-to-face with his sister Ruthie. He stood there in shock, happy to see Ruthie, but still not understanding how this was possible. He could believe he was dreaming more easily than accept that Ruthie was standing right in front of him.

It was his sister Ruthie, and Jack wasn't dreaming. She had gone to a bank convention in Lisbon, Portugal, and had booked two weeks of vacation to begin immediately afterwards, once she was already in Europe. Ruthie knew Jack was in Sitges because Jack had given Pops the bank address in order to get his money. She had taken a chance that Jack was still there and had traveled nine hours by train from Lisbon to Sitges, hoping to find him when she arrived there. She went to the bank that Jack had used and had finagled his address somehow. A smile from a pretty lady went a lot further then. In today's world, she would have never been able to get Jack's address.

No one had eaten breakfast yet. Jack introduced Ruthie to Maria, who was delighted to meet Jack's sister and immediately wanted to take a picture with Ruthie. Jack motioned "Later," and Maria left, contented. Ruthie waited in the room, talking a mile a minute, while the boys showered and dressed.

They went for breakfast at their favourite spot, the place where their friend Sarah worked. Jack knew Ruthie would enjoy eating by the ocean. Who wouldn't?

Sarah saw them walk in and came right over to their table to say hi and take their orders. The boys were just like the locals, having been in Sitges for over a month already, so they had their favourite table. Jack introduced Ruthie to Sarah, and they ordered their breakfast. They started off with three fruit salads. They were doing a lot of talking, so it was nice to have something in front of them that didn't get cold. Ruthie told them that they had missed a whopper of a snowstorm, which was pretty normal for a harsh Montreal winter. This storm had covered cars completely and made it impossible for anyone to go anywhere until they had cleared the roads. That took three days. Pops had to sleep on his desk at work for three days until he was able to get out and go home. Jack and Joe smiled, happy they didn't have to deal with that weather. Naturally, they felt bad for Pops.

They sat there telling funny stories of what had happened to them along the way, and they laughed and laughed. Everyone had finished their fruit salads long ago. They called Sarah over and asked her if there was something they could order that they could all pick at. She suggested a dish of sliced tomatoes and fresh mozzarella cheese, with balsamic vinegar, a little olive oil, and basil leaves. They ordered that, with a couple of orders of garlic bread.

Jack was so happy to see Ruthie. It brought him back home a little bit, and for the first time since he was on the road, he felt homesick. It doesn't matter how great a time you're having; there is no place like home. Living out of a suitcase was taking its toll.

Ruthie didn't want to change into her bathing suit and sit by the ocean, even though she was a big fan of suntanning. Jack knew right away what Ruthie wanted to do. She was leaving for home the next day, so she wanted to shop. Shopping for clothes was her love, but poking her nose into every curio shop along the way was what really appealed to Ruthie. So they headed downtown and had an amazing afternoon. After Ruthie had enough, the boys helped her carry all her bags to her room. She bought two beautiful tops that looked great on her, some souvenirs for people back home, and even some baked goods to pig out on. Ruthie never passed a bakery without checking it out. On the way to her room, they passed the bakery that the boys frequented, and in she went.

Ruthie was leaving the next morning for Barcelona. They left her room with plans of meeting a couple of hours later, for the "last supper." Her room was just five minutes away. Everything was just a few minutes away.

When the boys got back to their room, they sat around talking, and Ruthie came up in conversation. They couldn't believe that a single lady, traveling by herself, could have literally tracked down her brother in a foreign country. She had to get from Lisbon to Sitges, knowing it was very possible Jack was no longer in Sitges. She had to illegally obtain Jack's address from the bank and then find the place and show up at his door. Jack was glad Ruthie was his friend and not his enemy.

Jack told Joe that seeing his sister had made him think of going home. Joe felt the exact same way. It was time to plan to go home. They still had miles left to go on their journey. Jack asked Joe if he wanted to go back to Munich and see Gina and Lola on the way home. Those two girls were such good people and so much fun to be around. It was just meant to be that the boys left the beer hall with the two prettiest girls in town last time they were in Munich. It couldn't have been better planned. Jack and Joe talked about their good fortune all the time.

Joe immediately agreed. All four of them got along really well. He relished getting together with Lola again. So the plan became to head towards Munich and surprise Gina and Lola and then to fly home from Munich. There was a lot of "in between," but at least they had a rough start of a plan.

They arrived at Ruthie's room right on time. She was ready, so off they went. Earlier that afternoon, Ruthie had seen an Italian restaurant downtown that provided entertainment. Everyone was good with Italian.

The meal was excellent. The entertainment was very good also. A guy sang and played the guitar. Ruthie wasn't much of a drinker, so they all ordered a fancy drink that came with a sun umbrella straw. They toasted to health and happy trails. They toasted to family and friends.

Ruthie was leaving the next morning. She was taking the train to Barcelona and then flying directly from Barcelona to Montreal. She had to go back to work. They arranged to meet for breakfast the next day and then go to the train station together. It was a perfect time to tell Ruthie that they were planning to return to Montreal too. They would probably be home within the month, God willing. Ruthie, of course, had no idea about the hash. It was imperative not to confide in anyone, not even sister Ruthie. Not because Ruthie was untrustworthy by any means. It was about a facial expression or a comment made at the wrong time. It was best that nobody knew their secret.

On the way back to Ruthie's room, they stopped in at the bakery. They each bought a pastry to bring back to their room. They dropped Ruthie off at her room and then headed straight to theirs. They had to get up early to say adios to Ruthie.

The next morning, as promised, the boys were at Ruthie's room at seven. Again, she was ready and waiting for them. They went back to the restaurant where Sarah worked. Ruthie thought they made a great bacon, lettuce, and tomato sandwich. They substituted something else for the mayonnaise that Ruthie liked. The boys thought they did a pretty good job with everything, after eating quite a few items on the menu. Jack and Sarah were kindred spirits from day 1. He loved Sarah. The restaurant was by the ocean, and the food was great. Three good reasons to go there every day at least once and see Sarah.

They finished breakfast and went off to the train station. There were the appropriate hugs and kisses, and the train for Barcelona rolled in. Ruthie boarded, and poof, she was gone.

Life Goes On

After Ruthie left, the boys continued their routine of going to the ocean every day. They swam and exercised and got darker and darker. It was ocean in the morning and early afternoon, until they were ready for lunch and the bar at night. Tough life.

By this time, the boys were bringing friends back to their room quite often. Maria seemed to enjoy when they did. On one occasion, Maria insisted on taking a picture of these two girls from Venezuela that the boys had brought home from the bar. They all agreed to take the picture, the girls not knowing what was going on, but didn't care.

As always in this picture-taking situation, Maria hollered down to Tally, and she came running upstairs, on cue. Maria had everybody where she wanted them, so Tally made a pretend smile, then everyone smiled, and Tally took the picture. That picture of those two girls whose names Jack couldn't remember, and Maria with the boys, stood front and centre on Maria's wall.

When Jack got up the next morning, he told Joe he was ready to roll. Joe said that Jack had read his mind. That was why Jack and Joe were best friends. They often saw things in a similar way. They figured they would run around and say their goodbyes and leave the next morning. They had been in Sitges for over two months and had made a lot of friends, so maybe it was a little unrealistic to leave so fast, but they would try. Top on Jack's list was to say goodbye to Sarah. And of course, they had to tell Maria that they were leaving. Jack ran out to the florist and purchased a dozen red carnations, Spain's national flower. He brought them back to the room, and then he and Joe went to see Maria. Using only sign language and the odd Spanish word, they conveyed to Maria that they were going home and handed Maria the flowers. Maria's eyes welled up with tears, and they all hugged. They were genuinely sad to say goodbye to Maria. From that first night she saved their bacon, she was always a delight.

They went to the ocean for the last time. The regulars they had befriended were all there, so they were able to say adios to everybody. Jack had his address book in his bag, and everyone passed it around, scribbling their information into it. The book was almost full, holding dozens of names and addresses from all over the world.

The boys worked on their tans until lunchtime. They were super dark. They went for lunch, for the last time, at Sarah's restaurant. Sarah was extremely busy but saw them come in and went right over to say hi. Jack knew she was busy, and he couldn't tell her he was leaving as she whizzed by his table. So he asked her when she was finished working. She mouthed 2:00 p.m. Jack told Sarah he would wait, not to leave. She nodded okay.

The boys didn't have to wait long before Sarah finished work, and she joined them at their table, bringing three beers with her. There was small talk, and then Jack blurted out that they were leaving. Sarah was caught off guard and showed mixed feelings of happiness, trying to share in the boys' excitement and sadness that they were leaving.

Since Sarah was finished work, Jack asked her if she wanted to hang out together for the evening. Jack and Sarah had never hooked up, for some reason. They really liked each other, but it was never meant to be. If they had hooked up at the beginning, life in Sitges would have been very different for Jack. Sarah accepted the invitation, and off the three of them went. They took Sarah to their bar so they could say goodbye to the afternoon crowd, many of whom seldom showed up at night. The regulars were all there hustling the tourists, trying to sell them tours of the old town of Sitges. Sarah knew a lot of the regulars and was very much a people person, anyway, so Jack didn't have to worry about her having fun. The bar set up a backgammon tournament that was scheduled to start in an hour. All entries had to pay three dollars to sign up. If you lost a game, you were out. The person left standing got 90 percent of the pot, and the bar got the other 10 percent. There were thirty people already on the list. Anything could happen in backgammon, and luck was a huge factor. But Jack paid nine dollars and signed all three of them up. Sarah had mentioned to Jack weeks ago that she had played in a backgammon tournament and almost won. Jack and Joe were no slouches. They knew their way around a backgammon board too. So they had a chance to possibly walk away with the pot.

Jack told Joe and Sarah not to go anywhere, as he had just signed them both up for the backgammon tournament. They were pleasantly surprised and excited to play.

They made a pact that the winner would divide the winnings into three. Everyone agreed with high fives.

The tournament was fantastic. A number of people signed up but were half in the bag when it was their turn and were easily eliminated. A whole bunch of others were having too much fun with the ladies and forfeited their games and the three dollars that went with it. A lot of people had signed up; in total, there were thirty-four. That was roughly ten dollars to the bar, and thirty dollars for each of them. They couldn't chicken count, because there were still a few rounders in the mix of people left in the tournament that they had to beat.

Jack, Joe, and Sarah ended up playing one another at the end. They actually finished, one, two, three. Jack won, Sarah came in second, and Joe finished third. Being the winner, Jack had to buy everyone in the tournament a beer. Luckily, most of the people in the tournament had already left the bar. They were so high on the win. Of course, winning the money was great, but winning the tournament was exhilarating with all its bragging rights.

Jack went to Joe and told him that he wanted to spend his last night with Sarah, even though he hadn't asked her yet. Joe said he was going to stick around the bar and talk to stragglers and then head home. He would see Jack back at the ranch in the morning. Jack said thanks.

Jack went directly to Sarah, who was waiting for him on a barstool. He asked her what she wanted to do now. She said she was tired; she had to go to work at six in the morning. She added that she had a wonderful evening. It was now or never. Jack confessed to Sarah that he was determined to spend time with her on his last night in Sitges.

She said he was welcome to come along with her if he liked. Jack told her that he would really like that. Sarah laughed.

The next morning, Sarah started getting ready for work at about five. Jack had to jump in the shower right after her in order to be ready to leave with her. They said their final goodbyes at the door, and after a kiss and a squeeze of a hug, they went their separate ways. Sarah was gone, probably forever, but there was a big star beside her name in Jack's little address book.

Jack hustled back to their room at Maria's house. He hadn't gotten any sleep the night before, so he was looking forward to a little lay-down before heading back to Barcelona. No such luck. Joe was up, and getting any sleep was not part of the agenda. He was waiting, all packed up, ready to go.

He had gotten the hash out of the vent and had replaced the cover perfectly. Jack's two sacks of hash were laid out on the bed, waiting for him to return and load up. The thought of traveling with boots, let alone filling the boots with hash, was almost depressing. For over two months, they had been wearing sandals, shorts, and a T-shirt.

Joe asked Jack about the night before with Sarah. Jack just answered, "Wonderful."

They had to pack up, say goodbye to Maria one last time, and then get some breakfast. Jack immediately thought of Sarah. If they got a relatively early train to Barcelona, they could stay on Las Ramblas for a day or have the option of carrying on through Spanish customs and into France and French customs. Just get it all out of the way.

They decided to keep on going straight through to Paris. It's amazing how powerful the mind is. Once your mind is set on going home, it is easy to rationalize that you had your fun; it's time to go home. And you have little desire to go back to Las Ramblas, a dream come true. Dreams one day, memories the next. If they hadn't done it, though, they might have been leaving with some regrets.

Off they went. Maria was waiting by the front door, of course, for a final hug and another emotional goodbye, all in sign language. In the several months they spent under Maria's roof, they communicated perfectly with her, never speaking one word of English.

Next, they went for breakfast at Sarah's place. She laughed when she saw them walk in. She ushered them to their usual table on the patio by the ocean, scooted off, and returned with two coffees. They told her that they were all packed up and on their way to the train station but wanted to have a good breakfast under their belts.

As they ate breakfast, they said goodbye to the ocean. They both felt good about their experience in Sitges. Jack got up to go to the washroom and purposely crossed paths with Sarah. He bent over and whispered in her ear that he would never forget her. She looked at Jack with her big brown eyes and mouthed, "Me too."

As they sat waiting for their train, they made their mental goodbyes to a place they had enjoyed for over two months. They sat there in their jeans and boots and realized they were back on the road again.

Homeward Bound

The train for Barcelona arrived, and they were off. The train ride was quick, about forty minutes. They had already decided to go straight through to Paris. They grabbed an expensive burger and fries and sat down on a patio to eat it. It was a lot different sitting on a patio in jeans and boots. They were hungry and wolfed down their meals, hardly taking a breath. Jack took back anything he'd said about the price; the burger was delicious. It might have been the best burger Jack had ever eaten, but they were really hungry.

Now that they were fed and watered, it was on to Paris. Jack was happy they had decided to keep on trucking. Staying another night in Barcelona would have been going backwards, as tempting as it was.

They went to the train station and purchased two tickets to Paris. First, they had to clear Spanish customs. The boys were always hassled by customs agents and the police because of their long hair. Now, it seemed different. The customs guys looked at their tans and saw all the stamps in their passports and just waved them through. Not even a luggage check. No nothing. Just stamp, stamp, and they were off. It was all very routine.

They arrived in Paris. They had decided to continue on to Munich, but first, their train was stopping in Zurich, Switzerland. So far, they had only paid to go to Paris. They had to get off that train and purchase a ticket on another train going directly to Munich.

As they waited in the line to disembark with the other passengers, Jack saw a policeman with a dog on the platform. A dog was their ultimate threat. There was no way they could get off the train if they had to walk by that dog. Jack motioned to Joe with his chin towards the window, and Joe instantly made the connection. Jack told Joe to turn around calmly and to go back to their seats. Jack followed.

They sat in their seats as if they had a ticket to proceed to Zurich. Hopefully, the train would pull out, and when the conductor came around to see their ticket, they would plead ignorance and offer to pay then. They didn't care what would happen. The only thing they knew for sure was that they couldn't get off that train, no matter what.

The train did pull out. Jack and Joe were so relieved. They were on their way to Zurich, Switzerland, and then on to Munich. That meant another unexpected customs to be cleared.

They waited in their seats, rehearsing their story for the conductor, who would eventually show up to see the tickets they didn't have. It was nerve-racking waiting for the hammer to fall. It was time to keep the faith, again.

The conductor finally entered their car and began checking the passengers' tickets as he walked down the aisle. When he got to Jack and Joe, Jack nonchalantly produced their tickets from Barcelona to Paris. The conductor waved his index finger from side to side, indicating that they needed to purchase another ticket to Zurich and Munich. Jack told the conductor that no one told them they had to pay in Paris. Someone told them they were supposed to pay the conductor en route to Zurich, Jack lied. The conductor seemed to ignore Jack's explanation, did some calculations, and presented the boys with the additional cost. Jack paid the conductor the extra cost for both tickets, and it was over. It was that easy. Jack thanked the conductor for rectifying the situation. Again, he waved his finger from side to side: "Don't do this again."

They arrived in Zurich. Once all the passengers going to Zurich had disembarked, the Swiss customs boarded the train and began checking the passports of all the passengers still on the train. When they got to Jack, he presented his passport. The Swiss custom guy leafed through the many pages full of stamps, raised his eyes to stare at Jack, and then the same intense look at Joe, and without a word moved on to the next passenger. Boom, boom, boom.

The train started to move. They were on their way to Munich, finally. They had been on trains since they had left Sitges. There had been several close calls, the dog being the worst. That had caused a lot of stress. But it was all over. The boys nodded out in their seats and fell asleep for the remainder of the trip. They needed it.

They arrived in Munich. The boys had slept for several hours, so they were good to go. But they were hungry. They decided to get a bite to eat and then phone the girls. Maybe the girls had dates with a couple of other guys. After all, they had no idea that the boys would ever return to see them. And they were two ladies who easily attracted many guys.

First, they had to clear German customs. Again, it was a formality. They said, "Passport, please." Jack handed the customs guy his passport. The customs guy thumbed through the pages of stamps. He asked how long the boys would be staying in Germany. Jack replied, "Two days." The customs guy actually told Jack to enjoy his stay. Jack thanked him and went through. Joe breezed through as well.

They found a place that served a real German breakfast. They started off with a glass of orange juice and a coffee. Then they had rye toast with butter and jam. A basket of other breads and different rolls came to the table, with a plate of thinly cut slices of ham, salami, and two kinds of cheese. Jack and Joe made several sandwiches until they were stuffed. Jack couldn't remember the last time he had eaten such a huge breakfast, but it seemed to go down really well. Joe didn't have a problem finishing his breakfast either. The last time they had eaten was in Sitges, when they had breakfast at Sarah's restaurant. They had bought a few pieces of fruit for the road, but that was all they had had to eat in the past twenty-four hours.

Once breakfast was done, it was still really early, but they figured they could catch the girls before they went to work. They called Gina and Lola. They were really nervous that their little rendezvous wouldn't work out and they had come to Munich for nothing. After a few agonizing rings, Gina answered. She was astounded that the boys had returned to see them. Lola was there too. Gina told Jack to stop talking and get over to their place tout de suite. The girls were thrilled, but so were Jack and Joe. Jack thought they had been really lucky to catch the girls before they left for work but hadn't realized it was Saturday. They had the whole weekend with the girls before flying home on Monday. Just one more border to cross after they left Germany.

They grabbed their things and hustled over to the girls' apartment. There were lots of hugs and kisses when the boys arrived. The boys really poured it on how they couldn't go home without seeing Gina and Lola again. The girls just ate that up, but the guys really meant it. They sat around talking and laughing. Lola brought out some snacks, but the boys couldn't eat a mouthful, still weighed down with their huge breakfast. They began discussing what they would do that evening. Gina suggested they go back to the beer hall where they had met. Everyone agreed enthusiastically. It was a really fun place to go. Last time Jack was at the beer hall, he'd had a blast. This time would be even better, because they already knew Gina and Lola. They ended up staying at the apartment, fooling around, until it was time to go out to the beer hall.

Before they left for the evening, Jack wanted to book the flight back home to Montreal. They hit a great deal moneywise, which was wonderful. The drawback was that it wasn't a direct flight. It landed in New York first and then went on to Montreal. That meant going through American customs also. Reluctantly, they went ahead and booked the flight. They were going home.

The two couples left for the beer hall. When they walked in, the oompah band was playing as inebriated patrons sang and clapped their hands. It was like walking into an instant good time. The atmosphere was electrifying. Jack found some place at a table suitable for the four of them, and they all sat down. They ordered a pitcher of beer, which took less than a minute to materialize. When all the mugs were filled, Jack made a toast that simply said, "Be here now." Everyone clinked their mugs and drank. Joe lifted his mug and toasted to good friends, and everyone clinked their mugs again. Gina lifted her mug and toasted to happiness. Again, they all clinked their mugs and drank up. Lola didn't want to be left out, so she lifted her mug and toasted to love. Everyone looked at her weirdly, but they all drank to love.

After filling four mugs, there wasn't much beer left, so they ordered another pitcher. They were making friends with everyone around them. That's what people do at a beer hall, everyone feeling no pain. Jack and Joe noticed all the guys were majorly flirting with their gals, even in front of their wife or girlfriend. The boys knew it was harmless fun, certain they were the ones going home with the girls. So they turned a blind eye to the lovesick looks and innuendos. The girls were enjoying the attention and poured it on a bit, knowing they were safe with Jack and Joe there.

In the wee hours of the morning, people started to leave, and so did they. Lola was hungry, so they stopped at a pizza place and shared a medium pepperoni pizza before heading back to the apartment. Once there, both couples headed off to bed.

The next morning, their last full day with Gina and Lola, everyone got up late. It became a day of fighting grogginess and just sitting around, eating, drinking, and occasionally disappearing in couples for a nap. Jack knew that day with Gina and Lola would always be remembered as their last day in Europe.

Montreal or Bust

Lola and Gina were up early in order to get ready for work. Jack and Joe got up early as well so they could shower and get ready to leave with the girls. They had to load their boots up without the girls noticing. The girls were busy getting ready, so it was no sweat for Jack and Joe to load up properly, making sure their ankles were protected. This was it, no room for any slipups.

Everyone was ready, and it was only a final three-second kiss and the girls left for work and the boys took off for the airport. All goodbyes and promises were made long before they stood by the front door.

At the airport, they went straight to the Lufthansa counter, arranged everything, and headed towards the departure gate and German customs. Airport security was much tougher than train security. The boys were being very careful and in tune with their surroundings at all times. They were on the last leg of their journey, and every move was calculated.

They waited in the line to see the customs agent. When it was their turn, they approached the agent together. Those agents never smiled at you, only to one another. He asked them for their passports. He opened Jack's passport to the first page, the photo page, raised his eyes to look at Jack's face, and put the passport down. Then, he picked up Joe's passport and did the same thing. Then, out came the stamp, and boom, boom, they were off to New York. When they landed in New York, they would have to go through American customs. There would definitely be a luggage search, but hopefully, nothing more. Jack was praying there wouldn't be any dogs. If there were dogs, they decided they would hide in the washroom for hours if they had to.

The flight took between eight and nine hours. They were landing in New York at approximately 8:00 p.m., Munich time. There was a six-hour time difference, making it 2:00 p.m. when they landed in New York.

When they landed, everyone had to go through a luggage inspection. Joe's bag was first to be checked. As the customs guy felt his way through Joe's clothes, he stared at Joe's eyes, searching for fear or nervousness. He saw none. Jack didn't notice any of the other passengers getting as thorough an inspection as they did.

The agent came across a ceremonial dagger that each of them had purchased in the Middle East. He took it out of its sheath and examined the curved blade. He commented on what a beautiful dagger it was. Joe agreed and added that he was pleased he had bought it. The customs guy moved on, and next it was Jack's turn. The officer looked deep into Jack's eyes and saw no fear or nervousness with Jack either. The customs guy found Jack's dagger too. He pulled it out of its sheath and inspected the blade. Jack had used that dagger to chop up the hash when they had made that production line for rolling joints when they were in Sitges. Jack had cleaned his dagger carefully but never imagined that this customs guy would have the blade up to his nose as he checked out the intricate carving of the handle. Jack stood there hoping he wouldn't detect any signs of hash or even any whiff of their contraband. Satisfied, the agent put the dagger back into its sheath and moved Jack along. That was the hard part. Now they would have to answer some routine questions and get their passports stamped.

The passport guy checked their pictures and quickly leafed through the pages of stamps. He asked them where they were going. They answered they were going home to Montreal after ten months of traveling. After hearing that, the agent waved the boys through.

Now that they had made it through American customs, they only had the Canadian customs to deal with before they were home free. The safest place for them to be right then was at their departure gate. They only had to wait twenty minutes before they were able to board their plane to Montreal. An hour later, they would be home.

The flight to Montreal was so short the landing in New York and the taking off to Montreal, and now the landing in Montreal, caused Jack to get a severe earache. He couldn't wait to get off the plane. He sucked on a hard candy and swallowed a lot, but his head felt like it was about to explode. It was brutal. The biggest moment of his life was about to occur, and Jack couldn't see straight. He was in absolute agony. He confided in Joe but didn't inform the stewardess of the grave situation. Joe looked panicked and told Jack to hang on, not having any idea how he could help him. And hang on Jack did, without a whimper. He was hopeful that the problem would improve once the plane landed, and that was only minutes away. If it didn't subside, he definitely had to get some help.

Finally, in what seemed like a very long time to Jack, the plane landed. All the passengers clapped as the landing gear touched down on the runway. Once they were on *terra firma*, Jack's intense pain started to wane, then disappeared completely. With all the flying Jack had done, he had never experienced the excruciating pain he had just endured on that flight. It was scary.

Passengers began to disembark. Those needing to connect with another flight went one way, and everyone else lined up to clear Canadian customs. The line chugged along slowly. Jack and Joe were filled with anxiety as they neared the finish line. Then, they were next. Again, they approached the customs guy together. The agent asked for their passports, matched their identities with their pictures, thumbed through the pages of stamps as all the agents had done before, and smiled as he welcomed them home. It was unbelievable. It was all in the vibes you gave off.

Home Sweet Home

They were home. As they walked toward the terminal exit, they could hardly contain themselves. It was like they were walking on clouds. Even though they were bursting with excitement and felt pretty smug about themselves, they kept their cool until they were in the cab going home. They got into the taxi, and the moment the driver stepped on the gas, they high-fived each other and hugged, a robust hug. Jack caught the cabby in the rearview mirror, staring at them hugging, wondering what was going on. They had done it. They had done a really reckless thing, and Lady Luck had shown up at the right times, but they were smart like foxes when dealing with customs and the police. Not many could have pulled it off. It was hard to show no fear when you were really scared to death inside.

They decided to go to Joe's parents' house first. Jack could phone Pops and Ruthie from there and surprise them. Joe had last seen his parents in France, and Jack had seen Ruthie in Spain, but Jack hadn't seen Pops in over nine months.

When they showed up unannounced at Joe's house, everyone was home and happy to see them, especially Joe's younger sister, who idolized him. They had to excuse themselves immediately after the initial greeting, complaining they both had to use the washroom really badly. Everyone understood. They didn't really have to go to the washroom; they just wanted to remove the hash from their boots for the very last time. It was such a relief to not have to worry about hiding the hash anymore. The boys were thrilled to be home.

They socialized with Joe's parents for a while. Jack was still very appreciative of their generosity in Paris. The tour they had brought the boys on was first-class. Joe's younger sister was all over them, asking zillions of questions about their travels. Jack left Joe to answer her questions and excused himself. He took the opportunity to phone Pops.

Pops was glad to hear that his son had made it home in one piece. He, of course, didn't know how close it had come to a very different outcome. They made an arrangement to meet for supper at Murrays, Pop's favourite restaurant, the next evening. Jack said he would pick up Ruthie and see him at the restaurant at six.

Next, Jack phoned Ruthie. Ruthie was thrilled to hear from her brother. Jack told Ruthie he had just spoken to Pops and they had arranged to meet at Murrays at six the next evening. Jack said he would be getting his car out of storage in his friend's garage and he could pick her up at five. Ruthie said she would be waiting.

Joe told his sister some of the juiciest stories that he knew she would like. His sister smoked hash, so he told her about smoking hash in the Arab hash den. She was blown away. He told her stories about Israel and the kibbutz and meeting up with their cousin Ellie Weiss.

When Jack returned after making his phone calls, Joe took the opportunity to excuse himself, and the boys retreated to Joe's room. Joe had a phone in his room, so they called their good buddy Munroe. Munroe couldn't wait for them to return home. When he heard Jack's voice, he went nuts. Monroe insisted they get over to his apartment, tout de suite. They said they were on their way. Joe asked his dad if he could borrow the car for a few hours, and his dad had no objections. Jack piled the hash into a satchel, and they left for Munroe's place.

On the way there, a police car pulled up behind their car and put his flashing lights on. Joe pulled over. The cop came up to Joe's window and asked for his licence and registration. Joe gave him his papers. He explained that the car was his dad's car and he had borrowed it with his dad's permission. Jack tucked the satchel of hash between his legs a little farther so it was totally hidden behind his legs. The officer asked Joe to open the trunk. He did. The cop took his flashlight and checked every corner and told Joe to close it. The cop walked back to the car and took his flashlight and shone it on the back seat and then the floor. He did the same thing in the front seat, but the satchel was hidden from his view under Jack's legs. The cop told Joe that he had stopped him because one of his rear lights was burnt out. He told Joe that he would only give him a warning but he should tell his dad to get it fixed. Joe thanked the cop and assured him that he would.

They pulled away and continued on to Monroe's place. Jack told Joe that until the hash was totally sold, they had to stay vigilant and remain careful at all times. Joe agreed. They could have easily been busted minutes earlier.

That would have been tragic after everything they suffered through, just to get busted at home on a minor traffic stop.

When they got to Munroe's apartment, there was a lot of backslapping and bear hugs, everyone happy to see one another. At this point, nobody in the world knew what Jack and Joe had done. Jack put his hand in the satchel and began taking one sack of hash out at a time, placing them on the kitchen table. Munroe watched Jack with a bewildered look on his face. Jack knew Munroe didn't have any idea what was in the burlap sacks. Jack said, "It's hash. It's one kilo of hash." Munroe had never seen so much hash at one time. Then, it registered, and the questions started: "Where did you get it?" "How did you get it?"

Jack raised both his hands, stopping Munroe before his next question. Jack began to tell Munroe, whose mind was spinning a hundred miles an hour, the story of the hash. He began with first meeting that weasel Mohammed. He recounted everything that had happened, blow by blow. As Jack told the story, he almost couldn't believe the story himself. Munroe just sat there, very quietly, his mouth open in disbelief.

When Jack got to the part about being driven to the bus terminal in a limousine and dropped off without a plan in place, Munroe kept saying, "So what did you do?"

The next part, of course, was the biggest mistake the cops had made. Jack was sharp enough to say that he couldn't understand Hebrew when approached by Number 1 Cop. It turned out to be their saviour. All the cops left. If they had told a couple of cops to remain and guard the boys while they went to get Number 2 Cop, it would have been just like Mohammed and Number 2 Cop had planned.

Jack and Joe would likely be sitting in a Middle Eastern jail if they were still able to sit down. Jack told Munroe that Mohammed and Number 2 Cop were probably still trying to figure out where the hell the hash went. Munroe was in awe, his mouth still hanging open.

Jack and Joe both told Munroe nutshell versions of what they had seen and done. It was an exciting mishmash of stories. Then, the important topic of money came up. In Montreal, an ounce of hash was selling for seventy to eighty dollars. They had about thirty-five ounces. And it was really good shit too. They had smoked some along the way, and they wanted to keep a few ounces for themselves, so they decided that they would sell thirty ounces. That would bring in a mere twenty-four hundred dollars for their seventy-five-dollar investment. Joe questioned why they had to sell it by the ounce. It would take a lot longer, but they could sell the hash by the gram and double their return easily. Sell to people who couldn't afford to buy an ounce, those who bought a couple of grams at a time. The downfall was, they would have to deal with more people, which meant higher risk. But that was what they did, anyway.

It was a lot of work to cut up the hash into grams. It took almost four months to sell everything, but they finally sold all that they wanted to. Between the deals they gave friends and the hash they smoked themselves, they ended up with five thousand dollars, each getting twenty-five hundred. Joe paid his old man back the money he had borrowed, and Jack paid back Pops. They both had enough money left to easily finance another adventure.

Joe had always wanted to go to Marrakech, Morocco, and ride the Marrakech Express, a train made famous in the song by Crosby, Stills, Nash & Young. Nash had traveled from Casablanca to Marrakech in 1966 and had written the song "Marrakech Express" based on his experience on that train. That song hit a chord with Joe.

Joe left for Morocco shortly after they split up the money. Joe wanted to travel with Jack, but Jack had gotten an offer to go back to the Middle East. So Joe went to Morocco by himself.

Jack had an offer from Leon, the dude going out with his sister Ruthie. Leon was a very enterprising young man who had just bought property in Israel. He wanted to open up a leather shop in the old city of Jaffa. He needed help and had asked Jack if he was interested. Jack thought that would be a great opportunity and had accepted. Leon had a location where busloads of tourists were dropped off to shop and then brought back to their hotels with bags bursting with all their purchases.

So Joe took off to Morocco by himself, and Jack went back to Israel by himself. When Jack said adios to Joe, Joe told Jack to give his regards to Number 2 Cop. They both laughed. And Joe was gone.

Jack left shortly afterwards. Ruthie and Pops drove Jack to the airport and said their goodbyes and wished him a safe trip. And Jack was off on another adventure.

CPSIA information can be obtained
at www.ICGtesting.com
Printed in the USA
BVHW090803071222
653622BV00001B/100